Fruitfulness

OXFORD STUDIES IN PHILOSOPHY OF SCIENCE

General Editor:
P. Kyle Stanford

Advisory Board
Anouk Barberousse (European Editor)
Robert W. Batterman
Jeremy Butterfield
Peter Galison
Philip Kitcher
James Woodward

Systemacity: The Nature of Science
Paul Hoyningen-Huene

Causation and Its Basis in Fundamental Physics
Douglas Kutach

Reconstructing Reality: Models, Mathematics, and Simulations
Margaret Morrison

The Ant Trap: Rebuilding the Foundations of the Social Sciences
Brian Epstein

Understanding Scientific Understanding
Henk de Regt

The Philosophy of Science: A Companion
Anouk Barberousse, Denis Bonnay, and Mikael Cozic

Calculated Surprises: The Philosophy of Computer Simulation
Johannes Lenhard

Chance in the World: A Skeptic's Guide to Objective Chance
Carl Hoefer

Brownian Motion and Molecular Reality: A Study in Theory-Mediated Measurement
George E. Smith and Raghav Seth

Causation with a Human Face: Normative Theory and Descriptive Psychology
James Woodward

Branching Space-Times: Theory and Applications
Nuel Belnap, Thomas Müller, and Tomasz Placek

The Nature of Physical Computation
Oron Shagrir

Perspectival Realism
Michela Massimi

Fruitfulness: Science, Metaphor, and the Puzzle of Promise
Chris Haufe

Fruitfulness

Science, Metaphor, and the Puzzle of Promise

CHRIS HAUFE

OXFORD
UNIVERSITY PRESS

OXFORD
UNIVERSITY PRESS

Oxford University Press is a department of the University of Oxford. It furthers
the University's objective of excellence in research, scholarship, and education
by publishing worldwide. Oxford is a registered trade mark of Oxford University
Press in the UK and certain other countries.

Published in the United States of America by Oxford University Press
198 Madison Avenue, New York, NY 10016, United States of America.

Library of Congress Cataloging-in-Publication Data
Names: Haufe, Chris, author.
Title: Fruitfulness : science, metaphor, and the puzzle of promise / Chris Haufe.
Description: New York, NY : Oxford University Press, [2024]. |
Series: Oxford studies in philosophy of science | Includes bibliographical references.
Identifiers: LCCN 2023048039 (print) | LCCN 2023048040 (ebook) |
ISBN 9780197666395 (hardback) | ISBN 9780197666418 (epub)
Subjects: LCSH: Science—Philosophy. | Research—Philosophy. |
Inquiry (Theory of knowledge) | Idea (Philosophy) | Virtue epistemology.
Classification: LCC Q175 .H35197 2024 (print) | LCC Q175 (ebook) |
DDC 501—dc23/eng/20240108
LC record available at https://lccn.loc.gov/2023048039

DOI: 10.1093/oso/9780197666395.001.0001

Printed by Integrated Books International, United States of America

To Maysan, for a fruitful union

Ideally, the pursuit of truth is said to be at the heart of the intellectual's business, but this credits his business too much and not quite enough. As with the pursuit of happiness, the pursuit of truth is itself gratifying whereas the consummation often turns out to be elusive. Truth captured loses its glamor; truths long known and widely believed have a way of turning false with time; easy truths are a bore, and too many of them become half-truths. Whatever the intellectual is too certain of, if he is healthily playful, he begins to find unsatisfactory. The meaning of his intellectual life lies not in the possession of truth but in the quest for new uncertainties.

—Richard Hofstadter, *Anti-intellectualism in American Life*

Contents

Acknowledgments

I gratefully acknowledge the effort of Parysa Mostajir, Matthew Slater, and Neil Williams, each of whom read the entire manuscript at two different stages and provided careful and constructive feedback in both cases. That's a huge debt; all writers should be so blessed as to have friends like these. I want to express particularly deep thanks to Cyrus Taylor, an extraordinary physical scientist and human being who patiently endured hour after hour of unrelenting interrogation on the practice and history of his science, providing insight, correction, illustration, and encouragement in due measure. Discussions with him and with many other colleagues helped me to significantly refine my thoughts and my claims about the nature of inquiry. These generous interlocutors include Aviva Rothman, Alan Rocke, Colin McLarty, Stacy McGaugh, Dana Tulodziecki, Bishr Haydar, John Huss, Jamie Tappenden, Marc Lange, Elisabeth Camp, Peter Vickers, Tim Lyons, Kyle Stanford, Paul Hoyningen-Heune, and Brad Wray. Brad kindly invited me to participate in a workshop on the sixtieth anniversary of the publication of *Structure of Scientific Revolutions*, where I received lots of helpful feedback from a number of superior intellects.

Commensurate debts are due to Peter Ohlin, Kyle Stanford, and an anonymous referee at Oxford University Press, who nudged me to shore up weaknesses and infelicities in my account, resulting in a much improved final manuscript.

Lastly, apologies to my family for my obsession with aptness. I believe aptness deserves far more attention than it has hitherto received. I suspect that they would say I've given it far more attention than anything deserves. But it *has* borne fruit.

1

In Pursuit of the Fruit

This study attempts to capture something which is central to the epistemology of rational inquiry but which I believe no existing epistemological framework can accommodate: the influence exercised by concerns about potential fruitfulness in the development of scientific and mathematical knowledge. I came to appreciate the gravity of this problem in Fall 2011 when, after having taught Kuhn's *The Structure of Scientific Revolutions* several times, I finally decided to read it. What I found there was an intellectual achievement of bracing significance, one which I had been led to believe was bad philosophy of science but which, as I think many of us can now see, was actually attempting to hold a mirror up to philosophy of science into which its practitioners were simply not ready to look.

It occurred to me at that time that there was still an opportunity to take up a philosophical project that Kuhn's book had left for later generations—namely, to begin with the working hypothesis that theory or paradigm choice works pretty much the way Kuhn said it does, and then see whether some recognizable conception of epistemic rationality could be made to accommodate it. This led me to devote my attention almost exclusively to the history of science and mathematics, a body of knowledge with which I had almost no previous engagement. Focusing primarily on the very early developmental stages of ideas, I discovered a species of cognitive activity which my training in philosophy had convinced me could not exist. Although I had no idea how this activity could be conceived of as a variety of well-founded inference, its prevalence across the history of formal inquiry seemed sufficient to recommend it as the

Fruitfulness. Chris Haufe, Oxford University Press. © Oxford University Press 2024.
DOI: 10.1093/oso/9780197666395.003.0001

appropriate starting point for grappling with the epistemology of science and mathematics.

The history of science has since produced a decisive transformation in the image of science by which I was once possessed. Reading the history of science, and of mathematics, one is struck by how absolutely routine it is for a researcher to spend a lot of time—months, years, decades—getting essentially nowhere on a problem, and then suddenly see the form that the problem's solution will eventually take; more often than not, she will also see that the solution's form makes it suitable for a much wider range of application.

In this situation, we should expect nothing less than that the long-suffering researcher will cling to that solution's form like a raft in an endless, pitiless sea. She can see that the raft is hooked to an anchor, but she knows neither the length nor the elasticity of its tether. Inevitably she will endeavor to see how far she can travel on it, how wide of an area she can sweep out and still remain attached to something concrete. In the same way, a research problem's solution need not remain fixed to that specific problem. Rather, it is a vehicle—the *only available* vehicle—for going places in her field; a solved problem is also a *sample* of a solved problem. Given the need to pursue further concrete successes, she has a strong motivation to attempt to parlay that solution into a solution to another problem. How much sense would it make for her to jump off that raft and start a long, punishing swim with no guarantee of finding another raft? The natural response to this situation, and the one we see over and over again across the history of formal inquiry, is to try to stretch a problem-solving structure as far as it can possibly go. I believe that, until we give this basic fact about research the deference it warrants, our entire approach to the epistemology of science will remain fundamentally misguided.

This book is about fruitfulness, that seemingly intrinsic generative power that some ideas possess in abundance. What is fruitfulness? What makes some ideas especially fruitful? How do practitioners in mathematics and the natural sciences reliably select particularly fruitful conveyances for their investigations? And how does each of these questions bear on the astonishing power of formal inquiry? The purpose of this book is to answer these four questions, and to use those answers to solve a set of puzzles that pose different kinds of challenges to the cognitive authority of science. Each of these puzzles has its roots in well-established historical patterns of theory choice in science and mathematics, patterns which are, I think, very difficult to reconcile with the dominant epistemology of science, which tends to conceive of scientific knowledge as a set of justified (approximately) true beliefs about the natural world. However well-served analytic epistemology has been by this traditional conception, it has never fit comfortably on the actual structure of scientific practice. Recently, Hasok Chang (2022) has responded to this lack of fit by showing us how to understand notions like *reality* and *truth* in a way that preserves them as desiderata for an epistemology of science while respectfully declining to indulge in fantastical renditions of the nature of scientific inquiry. For anyone who thinks it important to retain these as epistemological desiderata, Chang's systematic pragmatist custom tailoring job provides a much needed scaffold. It is a monument to how maverick-y one can be while still coloring inside the lines.

I have always colored inside the lines. I even highlight the lines in bold by pressing down on the crayon really hard, just so everyone can see, yep, those are the lines, and I'm clearly coloring inside them. But there comes a time when you're sitting there, quietly coloring with your toddler, and you just have to stand up and shout, "These lines are bullshit!," so that everyone stops what they're doing and looks, and not a few tears are shed. That time is now.

The alternative perspective I develop over the course of this book grounds the epistemology of science in the ability of practitioners to select particularly apt frames for structuring inquiry. The selection of a research frame is a cognitive process, but one that is driven by prospective utility rather than past success. The structuring capacity provided by these frames gives rise to what I call *families of solvable problems*. On my account, fruitfulness is the propensity to produce these families.

The starting point for this project is a set of three basic challenges to the rationality of science raised by Kuhn's *Structure*. Here Kuhn famously drew attention to a number of aspects of theory choice which are historically well-established but whose relation to well-founded inference is somewhat mysterious. Take, for example, his observation that a research framework which eventually proves to be successful tends to quickly become very widespread in advance of any seemingly good reason for doing so, and even in advance of that eventual success. The apparent absence of traditional epistemic motivations in these instances contributes to an image of decision-making in research contexts that is considerably at odds with the kinds of philosophical resources with which we might hope to ground the cognitive authority of science. Even in its most developed form, Kuhn's account of theory/paradigm choice depicts it as, at best, an epistemically indistinctive phenomenon. And in that, it lends itself to a certain kind of abuse by those who seek to undermine science's status as epistemically special. For our dominant veritistic epistemology cannot countenance this phenomenon as a variety of well-founded inference.

The second challenge stems from Kuhn's attempt to take very seriously the likelihood that the content of our current theories about nature is historically contingent, and could easily have been otherwise. There is something epistemically destabilizing about the prospect of an alternative history in which we hold and venerate an entirely different picture of nature. Kuhn seemed to appreciate this destabilizing power, specifically for the way in which it challenged the rationale for esteeming our current scientific theories as correct and as the product of a species of inquiry that inevitably results in true beliefs.

The third challenge involves the well-established use of aesthetic criteria in the design and selection of these frameworks. There has often been a presumption in the philosophical literature that whatever explains the influence of aesthetic preferences on theory choice will have to discharge the burden of in some way linking a theory's possession of the preferred aesthetic properties to that theory's truth, or likelihood of being true. Although, as recently observed by Tappenden (2012, 211), this presumption has itself not been particularly fruitful, it is well-motivated, for behind it lies the recognition of two principles which should serve as constraints on philosophical models of formal inquiry:

(I) No factor would become a widespread feature of scientific and mathematical inference unless that factor improved the well-foundedness of inference.

(II) Improving the well-foundedness of inference means raising the probability that paradigmatically rational epistemic goals will be realized.

Under a traditional interpretation of science's epistemic goals, these principles work together to produce the suspicion that the aesthetic properties for which scientists and mathematicians often express a preference are in fact tied to truth in some way. For, if aesthetic considerations do characteristically govern scientific inferences in part, and if some kind of consideration could become a characteristic component of scientific inference only if it improved the well-foundedness of inference, *and* if improving the well-foundedness of an inference means raising the probability of its truth, then it must be the case that, when scientists and mathematicians discriminate among frameworks on the basis of possession of these properties, they do so because they are trying to raise the probability that the theory they end up choosing will be true.

I believe that (I) must be retained if there is to be any hope of preserving scientific inference as paradigmatically rational, and I hereby ratify it as a valid working assumption for philosophical models of science which take the preservation of science's special epistemic status as their solemn duty. I also regard (II) as sacrosanct: if the well-foundedness of a given inference is improved, then we're one step closer to realizing science's epistemic goals. And it's difficult to see how we could raise the probability of achieving science's *epistemic* goals without *any* improvement in the well-foundedness of inference.

But while there is wide agreement on the importance of (II), there is also considerable disagreement as to whether finding truths (even *significant* ones) is a good way of interpreting the dominant epistemic goal of science. Prominent alternative interpretations include the goal of "saving the phenomena" and the goal of increasing the ratio of solved-to-unsolved problems. I will have more to say about each of these proposals in what follows. But it will be important to keep in mind that each of these alternatives still takes science to be epistemically special, in the sense that scientific inference is in general, and not by accident, more well-founded in certain ways than nonscientific inference. Their differences lie principally in their understanding of what it is we acquire when we improve the well-foundedness of inference. Another way of describing the locus of disagreement is one where alternatives differ over precisely what sorts of epistemic improvements are made in cases where scientific choices are rational or progressive.

Examining the inference practices of mathematicians offers a way of rising above the fray by giving us insight into a kind of controlled environment, where practitioners typically must choose between a number of different theories whose truth is not in dispute. Since we know that concern for truth is not affecting theory choice in these contexts, we can move on to look at what other factors might be involved. Unless we regard these choices as arbitrary (which most of us do not), or unless we regard them as regularly governed by mathematicians' nonepistemic considerations (which most of us do not, hence (I)), we must find some way of explaining theory choice in mathematics that preserves the idea that mathematical progress is *epistemic* but which does not invoke truth-finding as part of the explanans. Given two true theories, how might we explain the rationality of a preference for one over the other?

We do not have the luxury (if it is a luxury) in the sciences of an abundance of competing theories of a given phenomenon, all of which we know to be true. What we do have, however, are a number of remarkable parallels between the sorts of properties scientists value in a theory, on the one hand, and what mathematicians value in a theory, on the other. We assume that the value of these properties in both mathematics and the sciences is epistemic, since (by (I)) they would not have become a standard feature of theory choice in these domains unless they contribute to the well-foundedness of inference. We also assume that mathematicians do not value these properties because of their association with a theory's truth. If the importance of these properties for making epistemically well-founded theory choices is insensitive to the properties' relation to the truth of a theory in mathematics, this would seem to justify the working hypothesis that these same properties improve the well-foundedness of theory choices in the natural sciences, whether or not those properties are associated with a scientific theory's truth.

The overarching thesis for which I will argue is that theory choices in the sciences are made on the basis of expected fecundity, and my proximate goal is to see how much of what we know about the mechanics of theory choice can be recovered by assuming that these choices are made on the basis of expected fecundity. To achieve this goal, I provide examples of how considerations of fruitfulness dominated particular theory choices in the history of science. While these examples are intended to provide some support for the claim that theory choices are made on the basis of expected fecundity, their primary purpose is to illustrate how the preference for fruitfulness has been pitted against more traditional epistemic desiderata. This historical juxtaposition between fruitfulness and other desiderata will be important for resisting the temptation to believe that scientists make theory choices on the basis of expected fecundity because fruitfulness is a guide to truth. In fact, something like the reverse appears to be the case—that is, when choices are made on the basis of presumed truth, it is because truth is seen as a guide to fruitfulness.

I begin in Chapter 2 by outlining the principal Kuhnian burden that my theory of fruitfulness hopes to discharge. I then examine the demands on an epistemology of fruitfulness from the perspective of epistemology more generally. One of the principal trials Kuhn faced as an intellectual was that he was an epistemological naturalist way before it was cool. Reading *Structure* sixty years later, it is apparent that Kuhn was caught between two intellectual aims that had not yet—perhaps *still* have not—been coaxed into peaceful coexistence. On the one hand, he exhibits a desire to treat scientific knowledge as exemplary of knowledge itself. On the other, he wanted to take seriously the contingency of the development of scientific knowledge. *Structure* provides no means for doing both, and his response to philosophers of science was always to further emphasize the accidental and capricious character of the understanding of nature that prevails at any given time. This tension lies at the heart of what I call the Puzzle of Promise. In Chapter 3, I begin to explore the sense of "future promise" routinely exhibited by practitioners in mathematics and the natural sciences, carving out the basic contours of the fully developed account I present in later chapters: solvable problems, exemplars, and research frames.

Chapter 4 marks the first of three chapters which together compose my positive account of how fruitful theories emerge and bear fruit. This chapter builds on the theory of exemplification as articulated by Goodman in *Languages of Art*. Goodman's reference-oriented theory gives us some purchase on developing a picture of how exemplars emerge in cognition in general and in rational inquiry in particular. An exemplar emerges when a stable cluster of features of experience stands out as particularly salient. These salient stable structures form the basis for the emergence of certain salience norms, which guide classification.

Chapter 5 then shows how research frames arise out of exemplary problem-solving attempts. Guided by Elisabeth Camp's systematic account of cognitive framing, I look in detail at several episodes in the history of science in which such exemplars have resulted in highly influential research frames, paying particular attention to the varying uses to which a single frame is often applied and the bases upon which it is selected. Rarely do exemplary problem-solving attempts stand as successful and enduring solutions to research problems. Rather, they constitute the basis for research frames that flourish over multiple generations of practitioners.

Chapter 6 describes how the attempt to squeeze every possible bit of utility out of a frame drives its use beyond the context in which it was designed to function, resulting in an increasingly figurative perspective on the frame's content. The increasingly figurative use of frames facilitates radically new perspectives—a familiar feature of the behavior of metaphor.

Frames are chosen on the basis of relative aptness. In Chapter 7, I argue that the judgment that a research frame is apt is of a piece with aptness judgments in the context of metaphor. Indeed, research frames function as a variety of metaphor. Like judgments of metaphor, aptness judgments in the context of research tend to be quick and highly uniform across practitioners. These judgments focus partly on accuracy, and partly on the cognitively congenial structuring that some frames provide for guiding inquiry. On this basis, I provide an alternate account of epistemic warrant in science that locates the well-foundedness of theory choice in perceptions of aptness. Chapter 8 examines aptness judgments in finer detail, focusing in particular on the role that aesthetic features of frames play in facilitating the structuring of inquiry. Aesthetic features of theory choice have long been observed as having a significant but epistemologically troublesome role in determining the course of the development of scientific and mathematical thought. I argue that these are expected features of theory choice if we understand theory choice to be made on the basis of aptness judgments.

I conclude in Chapter 9 with an examination of how the image of science and mathematics as a search for particularly apt frames bears on the threat to cognitive authority posed by the apparent historical contingency of the content of scientific and mathematical knowledge.

2

The Puzzle of Promise

> *We can also imagine the case where nothing at all occurred*
> *in B's mind except that he suddenly said "Now I know how*
> *to go on"—perhaps with a feeling of relief; and that he did*
> *in fact go on working out the series without using the for-*
> *mula. And in this case too we should say—in certain*
> *circumstances—that he did know how to go on.*
>
> —Wittgenstein, *Philosophical Investigations*, §179

Toward the end of *The Structure of Scientific Revolutions*, as the
epistemology of science spirals further and further out of cognitive
control, Kuhn describes the decisions of a natural scientist trying
to push inquiry forward during a period of paradigm change, a period in which "the priority of paradigms" no longer holds. In the
absence of guidance supplied by a track record of solved problems,
he observes, the scientific choices of the man "who embraces
a new paradigm at an early stage" are instead driven or at least
strongly influenced by "formal" and "aesthetic" criteria. Kuhn then
attempted to frame in the starkest possible terms the deep epistemic problems confronting someone in this position:

> the issue is which paradigm should in the future guide re-
> search on problems many of which neither competitor can yet
> claim to resolve completely. A decision between alternate ways
> of practicing science is called for, and in the circumstances that

Fruitfulness. Chris Haufe, Oxford University Press. © Oxford University Press 2024.
DOI: 10.1093/oso/9780197666395.003.0002

decision must be based less on past achievement than on *future promise*. (1962/2012, 156)

These early stages of inquiry present a serious challenge to the rationality of scientific inquiry: if decisions between alternate ways of practicing science must be based on *future promise*, how can those decisions be epistemically justified if not by previous success? Kuhn's answer to that question was—or at least *seemed*—deliberately unepistemic: "a decision of that kind can only be made on faith" (Kuhn 1962/2012, 157). It was not popular among those concerned with understanding and grounding the cognitive authority of science.

2.1. Retreat from Reason?

Was Kuhn just being provocative by invoking faith in relation to scientific decisions? As with so many claims in *Structure*, we want to say, "Partly yes, partly no." On the one hand, Kuhn is teasing out some of the threads of a metaphor that he used to characterize the transition to a new scientific paradigm: *conversion*. It is hard to believe that this particular metaphor was not chosen at least in part because of the religious connotations that it would surely have suggested to readers; it promotes an understanding of paradigm choice as being directly opposed to a certain widespread conception of scientific reasoning and its causal role in the development of scientific knowledge. That widespread conception holds that what makes natural science so powerful is the fact that practitioners proceed slowly and methodically through a series of well-defined incremental steps, the magnitude and direction of which are dictated univocally by the accumulated empirical evidence—or, at least, the steps *can* be univocally dictated by the accumulated empirical evidence, once we sort out the details about how evidence confirms theory. Kuhn, of course, was writing at a time in which recent developments in relevant branches of philosophy (Quine 1951; Wittgenstein 1953; Goodman 1954) had sowed the seeds for the inevitable destruction of this endearingly rosy epistemological paradise. But the full impact of the views of a few epistemologically woke philosophers on our understanding of scientific reasoning had yet to sink in; appearances suggest it has yet to sink in still. Philosophy of science has never been particularly fussed by the inconveniences of history, and Kuhn's contemporaries were no exception. He used *Structure* as a platform for accentuating the many epistemological deviancies that pervade the workaday world of the practicing scientist. What better way to draw our attention to those philosophically impious indulgences than by framing historically consequential paradigm changes in terms that were antagonistic to the use of reason?

While Kuhn may have enjoyed rankling philosophers of science, he was never one to rankle for rankling's sake. On and on he rankles, but always with purpose—and usually with discomfitingly compelling evidence. The reason that religious conversion seemed to him to be an apt metaphor for paradigm change is because it is thought to exemplify a process not grounded in rational deliberation. And the simple fact is that the history of science records several species of inference that really should not exist if scientific inquiry is a paradigmatically rational process by traditional lights—not a perfect one, but certainly good enough on its best day to teach us something deep about how knowledge works. No one was persuaded more than Kuhn of science's exemplary epistemological status: "How," he asks rhetorically, "could the history of science fail to serve as a source of phenomena to which theories of knowledge may legitimately be asked to apply?" (Kuhn 1962/2012, 9). But there are so many extraordinarily weird epistemic phenomena surrounding the decisions made in the course of inquiry. The closer one looks at the history of science, the more difficult it becomes to square traditional conceptions of scientific evidence and reasoning with the way that natural science actually works. Kuhn had made his choice: there are lots of instances—historically significant ones—in which reasoned deliberation seems not to affect the direction of scientific inquiry; often it cannot. He was not just being provocative. As far as we can tell, Augustus Kekulé actually did come up with the idea of the benzene ring after dreaming about a snake eating its own tail (Rocke 2010). Johannes Kepler's third law really was the outcome of his quest to discover the harmonic ratios exhibited in planetary music (Rothman 2017). Robert Millikan openly admitted to ignoring a significant number of oil drops in his Nobel Prize–winning study of the electron (Holton 1978). Duly considered, the history of science reflects epistemic goings-on fundamentally at odds with prevailing models of scientific reasoning. Faith often comes out looking pretty good by comparison.

Structure was an epistemologically radical book. But its claims about knowledge and reason proper are considerably more modest than those made a decade earlier by Quine, Goodman, and Wittgenstein. Why then was the philosophers' reaction to *Structure* so much stronger? Part of the reason, I think, had to do with his use of specific and concrete instances from the history of science as illustrations of his major claims. In contrast to his predecessors, whose heterodoxies subsisted primarily on imaginative abstracta which could be kept at some remove from the production of knowledge, Kuhn was describing things that had *actually happened in the history of science*, which seemed to somehow make the whole midcentury epistemological snafu more real. But the vituperation he endured also reflects our unspoken commitment to the principle that science is in some sense more sacred than knowledge. Attacking the foundations of knowledge is one thing. Describing science in a way that does not reflect the received view of what exemplary inferences must look like—well, that was not to be tolerated. His suggestion that the Temple of Reason was, like Notre Dame, just another monument to faith, did not exactly calm things down.

At the heart of this book is a puzzle about epistemic justification—the Puzzle of Promise—and in this chapter I describe the evolution of that puzzle. The Puzzle arises when one accepts the basic details of Kuhn's account of how paradigm choice works but rejects his epistemologically toxic interpretation of those details. Most philosophers of science reject his interpretation of those details; they more or less ignore the historical details themselves. Like most philosophers of science, I think much of what Kuhn says about the epistemology of theory/paradigm change, both in *Structure* and later, is simply wrong. The problem is that much of what he says about the practice of science and the considerations that dominate scientific decision-making appear to be right, as far as they go. In fact, as I describe shortly, there is a clear sense in which matters are considerably more dire than even Kuhn would have allowed— paradigms are routinely adopted before they can claim to have resolved any outstanding problems whatsoever. But there is one thing on which all parties seem to agree: the way in which paradigm choices are actually made is an embarrassment to traditional theories of scientific rationality.

The dominant theme of Kuhn's epistemology of science is the *future-directedness* of decisions made in a research context. This manner of thinking changes the calculus regarding the approach one feels compelled to employ in the pursuit of scientific knowledge, in ways that are deeply counterintuitive when viewed through an epistemological lens trained on the scientific confirmation of beliefs about nature. Kuhn's struggle throughout *Structure* is to invert that lens, affording us a perspective from which we could see what the *scientist* sees when she peers through it. The scientist must keep constantly in focus the question of how she is going to continue to engage productively in scientific investigation. From her vantage point, there is no end of inquiry. The aim of science is, ultimately, more science.

Reading *Structure*, you often feel that you are playing Wile E. Coyote to Kuhn's Road Runner. He had a remarkable talent for luring you blissfully off of an epistemological cliff. By the time you have the sense to check to see whether there is anything supporting scientific decision-making, you realize that you've merely been delaying your inevitable descent into oblivion. As the earlier quote indicates, this is a pitfall from which Kuhn himself was not immune. He, too, was a victim of the book's highly intoxicating combination of plausibility and radicalism. He understood that the practitioner's need to fixate on promise was too compelling to ignore, but that no available philosophical framework was capable of grounding the epistemic well-foundedness of its pursuit. Thus, his work becomes a very congenial venue for appreciating and exploring the contours of the Puzzle of Promise. Looking at its development over the course of his thinking is an interesting and instructive way to deepen one's grasp of the Puzzle of Promise. I follow this pattern of development across three separate dimensions. In section 2.2, I look at the epistemological questions raised by the Puzzle and how Kuhn's comments about them changed over the years in ways that are surprising, confusing, and deeply unsatisfying. In section 2.3, I discuss some of the pesky historical details that drove Kuhn into the seductive arms of epistemological heterodoxy. In section 2.4, I examine a small but critical refinement in the way Kuhn conceived of what it is that scientists acquire when they learn from exemplars such as Newton's *Principia* or Darwin's *Origin*. I close the chapter in section 2.5 with a refined version of the Puzzle.

2.2. Promise, "Postscript," and Post-"Postscript"

In 1970, Kuhn published a second edition of *Structure* which contained a significant postscript (hereafter "Postscript")[1]. The substance of "Postscript" is composed primarily of extensions, clarifications, and defenses of the central theses promoted in the

---

[1] Citations of "Postscript" refer to its reprinting in the 4th edition of *Structure* (2012), which begins on page 173.

first edition. Notably absent are any references to faith as a basis for paradigm choice.

Chastened both by well-deserved criticisms and by the palpable delight that many readers took in the idea that decisions of extraordinary epistemic import are not rationally grounded, Kuhn openly declared fealty to epistemic values such as "those usually listed by philosophers of science—accuracy, simplicity, fruitfulness, and the like"—values which, he conceded, could serve as "good reasons for choice" (Kuhn 1962/2012, 198). In other words, the choice of "which paradigm should in the future guide research" is emphatically not based on faith. It is—or at least *can* be—based on the kinds of reasons whose intimate association with rationality has long been acknowledged. Of all the attacks Kuhn received from philosophers of science, this is the only one where he seemed to offer an unconditional surrender. There's a part of me that has always admired Kuhn for this shameful capitulation. Those references to faith weren't just provocative—they were needlessly provocative. His choice of metaphor here was no small contributor to the bonanza of conscious efforts following *Structure* to weaken the epistemic credentials of scientific knowledge, a legacy which he painfully acknowledged but which could not have been altogether surprising (Kuhn 1993, 106). Witness for comparison the contrasting fortunes of the widely reviled thesis of incommensurability. He *never* yielded on this point; not one inch. He was still talking about "world changes" and "working in a new world" right up to the end (see, e.g., Kuhn 1993). He refined this notion over and over again throughout his career. And actually it only became more plausible as the years went by (Wray 2021, Part IV). Rolling over on the "faith" issue was not just Kuhn being Kuhn. He made an error, and he owned up to it. It was all most unbecoming of a philosopher.

And what, then, of promise? What are decisions about future promise based on, if not faith? There's no indication that Kuhn abandoned the idea that decisions between alternate ways of practicing science are geared toward the most promising among alternatives, or that he relented in his view that considerations of *"relative problem-solving ability"* were not up to the task of explaining why an incumbent paradigm is perforce rejected in favor of a mostly untested challenger (Kuhn 1962/2012, 156; emphasis added). The relatively uninspiring track record of successor paradigms is a philosophically embarrassing historical fact that constrains plausible solutions to the Puzzle of Promise; there was no reason to revisit it. Rather, it appears that he came to see the canonical epistemic values as some of the means by which practitioners set future promise in their sites.

Kuhn's views on the status and meaning of the epistemic values evolved after "Postscript" in a direction that I find surprising. There he is extremely cautious in his admission of their relevance to paradigm choice. Earlier I said that he openly declares his fealty to the canonical values; what he actually says is that, with respect to the notion that paradigm debates are about *persuasion* rather than *proof*:

> Nothing about that relatively familiar thesis implies either that there are no good reasons for being persuaded or that those reasons are not ultimately decisive for the group. Nor does it even imply that the reasons for choice are different from those usually listed by philosophers of science: accuracy, simplicity, fruitfulness, and the like. (Kuhn 1962/2012, 198)

Not exactly a ringing endorsement, I know. The most he can muster is that those reasons are not necessarily bad. Still, I call this progress, especially given his clear interest in *Structure* in depicting the decision between paradigms as a kind of choice to which notions of *good* and *bad* could not really apply.

From there, things begin to take an odd turn. Consider his famous essay, "Objectivity, Value Judgment, and Theory Choice," written a few short years later. Here Kuhn's embrace of the philosophers' epistemic values is unequivocal: they are "doubtless . . . among" the

criteria that govern paradigm/theory choice (Kuhn 1977, 325)—
quite a shift from that gratingly noncommittal concession in
"Postscript." There's more. This essay also contains the curious and
decidedly un-Kuhnian comment that these criteria constitute "the
canons that make science scientific" (Kuhn 1977, 324). There is a
lot one could say about this statement in terms of how it relates to
Structure's implicit conception of what "makes science scientific."
Let it be observed that, to the extent that *Structure* does contain a
conception of what makes science scientific, it has more to do with
research being paradigm-driven than it does with any criteria for
decision-making. Recall, for instance, this beloved passage:

> At various times all these schools made significant contributions
> to the body of concepts, phenomena, and techniques from which
> Newton drew the first nearly uniformly accepted paradigm for
> physical optics. Any definition of the scientist that excludes at
> least the more creative members of these various schools will ex-
> clude their modern successors as well. Those men were scientists.
> Yet anyone examining a survey of physical optics before Newton
> may well conclude that, though the field's practitioners were
> scientists, the net result of their activity was something less than
> science. Being able to take no common body of belief for granted,
> each writer on physical optics felt forced to build his field anew
> from its foundations. In doing so, his choice of supporting ob-
> servation and experiment was relatively free, for there was no
> standard set of methods or of phenomena that every optical
> writer felt forced to employ and explain. (Kuhn 1962/2012, 13)

Let's not dwell on the possible tensions with *Structure*. It's very difficult to pin down what, if any, notion of science Kuhn subscribes to
in *Structure*. That's not that surprising if one understands "science"
to refer to a historical lineage which takes on whatever properties
might be foisted upon it by the vicissitudes of history, as Kuhn presumably did; science is different things at different times. This is

precisely what makes that later comment so weird. The last thing that would occur to someone after reading *Structure* is, "Now *here's* a guy who thinks that paradigm/theory choice is decided on the basis of simplicity, accuracy, fruitfulness, and so on. Why, he even thinks that this is what makes science scientific!" That is not the message he intended to convey in *Structure*. He wanted to emphasize how distant paradigm choice is from the influence of rationality. By all accounts, he was successful.

Stunningly, his later work only deepens the semantic connection of the epistemic values to rationality. By the time we reach his career-capping "Afterwords," the canonical values have come to function as an *analysis* of rationality itself:

> the rationality of the standard list of criteria for evaluating scientific belief is obvious. Accuracy, precision, scope, simplicity, fruitfulness, consistency, and so on, simply are the criteria which puzzle solvers must weigh in deciding whether or not a given puzzle about the match between phenomena and belief has been solved. . . . [T]hey are the 'defining' characteristics of the solved puzzle. (Kuhn 1993, 338)

In case there was any lingering question of whether Kuhn was being hyperbolic in claiming that these values are what make science scientific, we now see that this idea was actually somewhat understated relative to the fully developed expression of his views. The influence of the canonical epistemic values isn't just what makes science scientific; it is what makes science a *rational* process, because to choose on the basis of these criteria is just *what it means* to choose on the basis of reason.

And now things go to a very epistemologically dark place. Kuhn continues the thought from earlier:

> It is for maximizing the precision with which, and the range within which, they apply that scientists are rewarded. To select a

law or theory which exemplified them less fully than an existing competitor would be self-defeating, and self-defeating action is the surest index of irrationality. (Kuhn 1993, 338)

Does everyone see what's going on here? The account of what makes scientific decisions rational that's being floated is:

1. Attempting to maximize the quantities for which one is rewarded is rational.
2. Scientists are rewarded for maximizing accuracy, simplicity, and so on.
3. Therefore, attempting to maximize accuracy, simplicity, and so on is rational.

Why is the imminent reward the thing that makes maximizing accuracy, simplicity, and so on rational? Why isn't it, you know, the fact that there is something *epistemically special* about those qualities? On this account, had scientists been rewarded for maximizing the complexity or hilariousness of theories, then making scientific decisions with an eye toward maximizing complexity or hilariousness would have been the epitome of rational choice. In other words, scientific inquiry would have qualified as rational *no matter which* values governed the distribution of rewards. As long as scientists are trying to maximize the quantities for which they're rewarded, they qualify as rational.

One is tempted to try to come to Kuhn's rescue by insisting that there's an equivocation in the use of the word "rationality" here—he begins by using "rationality" to refer to *theoretical* rationality and ends the thought with a reference to *practical* rationality. Yes, that checks out. So maybe he just made a mistake? I don't think so. The horrifying truth revealed by this "mistake," I think, is that he exploited the unfortunate overlap in the sounds we use to describe the system of norms for furthering our practical interests, on the one hand, and those we use to describe the system of norms governing the well-foundedness of inference, on the other, to deprive science of any epistemically distinctive rank. In the account of scientific rationality presented earlier, the explanation for why we associate accuracy, simplicity, and so on with the concept *rationality* in science is because adherence to those norms furthers the practical interests of scientists with respect to the incentive structure that, by an accident of history, happens to prevail in science. It has nothing to do with the contribution of those norms to well-founded inference or the growth of knowledge per se. Had scientists been rewarded for maximizing the hilariousness of theories, then maximizing accuracy, simplicity, and so on might easily "be self-defeating, and self-defeating actions is the surest index of irrationality." In other words, had modern science developed along a different historical trajectory, choosing theories/paradigms on the bases of accuracy, simplicity, and so on would have been *irrational* rather than rational. This fits nicely with his increasingly explicit post-*Structure* view that the canonical epistemic values are an *analysis* of what it means for scientific decisions to be rational. For Kuhn, there is no broader epistemological sense in which these norms are epistemically valuable. We call those norms elements of rational choice only because scientists are incentivized to pursue them rather than something else. Their association with reason is a mere byproduct of the contingent system of rewards that is peculiar to modern science.

Whether or not Kuhn was aware of it, though, there is a deeper question here, one of arguably more direct relevance to the question of epistemic justification, and one which, when properly stated, highlights how quite beside the point Kuhn's answer was— namely, why have intellectual traditions that have governed their choices in *this* way been *so successful*? What is it about *these criteria* that so generously rewards decisions based upon them? This is a question about the causal relationship between the set of choice criteria that define scientific rationality, on the one hand, and an associated history of scientific progress, on the other. Even if it is true, as Kuhn asserts, that accuracy, simplicity, and so on "are the 'defining' characteristics of a solved puzzle," could *anything* have been? *Could* hilariousness and complexity have ascended to the level of the solved puzzle's defining characteristics? There is no upper limit on complexity; maximizing a theory's complexity is not even a thing. More importantly, though, attempting to maximize the complexity of a theory does not serve the interests of knowledge. We associate the canonical choice criteria with rationality because we take it that there is something epistemically special about the choice criteria that actually prevail in science. Why? Because it is easy to imagine an alternative set of choice criteria resulting in a much less impressive alternate history of science. Kuhn's view suggests the opposite—viz., that the only thing driving the growth of knowledge across paradigms is the existence of a reward structure per se. This simply cannot be right. Any arbitrarily composed set of norms can constitute a reward structure. But many such sets will be antithetical to the growth of knowledge.

It is not difficult to find arguments in *Structure* that appeal to the causal role of this or that norm in contributing to the goals of paradigm-driven normal science. For instance, we can see from that perspective why, as he says, "[n]ovelty for its own sake is not a desideratum in the sciences" (Kuhn 1962/2012, 168) because "[n]ormal science does not aim at novelties of fact or theory and, when successful, finds none" (52); that textbooks' suppression of the history of science is pedagogically "unexceptionable" (139) given that textbooks are "vehicles for the perpetuation of normal science" (136); and that "[b]eing able to take no common body of belief for granted" makes normal science impossible (13). And on and on and on. The whole book is one gigantic causal model composed of causal processes governed by the cultural norms of science. Kuhn never suffered from a shortage of causal stories for the manifest power of scientific inquiry. He reveled in the presentation of surprising, heterodox, yet ultimately intuitive causal explanations for why science works so well. Then why is there no quarter given to the specific epistemic norms associated with rationality itself? Why are they not allowed to figure in a classic Kuhnian causal story like those about how paradigms make normal science possible or about how the rigor of scientific training results in the capacity to detect anomalies?

I think the answer is obvious: Kuhn never really abandoned his commitment to the epistemic arationality of paradigm/theory choice. "Faith" turned out to be an impolitic vehicle for expressing that commitment. He found that he could achieve essentially the same result by appealing to the canonical epistemic values while depriving scientific rationality of any real epistemic significance— that is, any meaning beyond that of just being the form that practical rationality takes in the context of science. To have the canonical values feature in a causal explanation of why paradigm choice curiously results in the eventual expansion of our knowledge would be to assign those values distinctive capacities that set them apart epistemically from other potential sets of values—sets which, for all he appears to think, would qualify as equally valuable epistemically so long as scientists were rewarded for maximizing them. It would, in addition, provide a foundation for elevating scientific inquiry above other approaches to understanding nature that did not subscribe to that set of values—or, at least, to a set of values with the capacity to cause an expansion of our knowledge commensurate with that of modern science.

Most importantly, though, any such explanation would implicate the canonical values in a causal process that results in "the actualization of . . . promise" (Kuhn 1962/2012, 24). And if we know that the maximization of these properties tends to result in the actualization of promise, then we can use a paradigm's possession of them as an index of its promise. As part of such a process, then, properties such as accuracy, simplicity, and so on would be able to serve as part of a rational foundation for the assessment of promise. In that case, judgments of future promise are not based on faith. Nor are they based on knowledge of what is likely to result in some juicy rewards. They are based on knowledge of what causes the actualization of promise. The factors that result in the actualization of promise do not occupy that causal role simply because of a historically contingent incentive structure, as Kuhn seems to think. Indeed, something like the reverse appears to be the case: the existence of the specific incentive structure peculiar to modern science—maximizing accuracy and the like—is explained by the fact that attempting to maximize those properties *causes* the growth of knowledge.

The essence of the Puzzle of Promise is the search for an alternative species of epistemic warrant in science that satisfies a variety of constraints imposed upon us by the history of science. The primary constraint driving the Puzzle is that choices about future promise cannot be based on a comparison between the problem-solving records of an incumbent paradigm and a challenger. According to Kuhn, challengers always lose that fight. Nevertheless, some challenger is eventually chosen because of its promise. As we've seen, "faith" is not an alternative species of epistemic warrant; it is designed to be the absence of warrant. Nor is Kuhn's version of what makes the canonical choice criteria "rational." But the sustained and, quite frankly, shocking success of science suggests that *some* species of epistemic warrant *must* attach to paradigm choice. Our analysis of Kuhn's amendments to his views about choice suggests that the alternative species of warrant we seek resides partly in an answer to the following question: *why does the attempt to maximize the historically important choice criteria result in the actualization of promise?* If we can explain why decisions governed by these specific choice criteria tend to result in the growth of knowledge, we will have articulated a conception of epistemic warrant that is both aimed at future promise *and* independent of the problem-solving record.

2.3. Shift Happens

Worried that readers might misinterpret his affirmation of the influence of the canonical choice criteria, Kuhn's discussion in "Postscript" pivoted to a reminder that the seeming unity of values governing paradigm/theory choice in the natural sciences disguises the fact that things like "simplicity" and "explanatory power" mean different things to different individuals: "the application of values is sometimes considerably affected by the features of individual personality and biography that differentiate members of the group" (Kuhn 1962/2012, 185). Thus, the fact that two practitioners subscribe to the same set of values does not imply that they must therefore make the same decision when it comes to choosing between paradigms. This same point is elegantly extended in his "Objectivity, Value Judgment, and Theory Choice."

Why did he feel the need to emphasize this? His fear appears to have been that some people would think that, were scientists to share epistemic values, it might be possible in principle to work out a decision procedure that would uniquely determine the optimum choice in a given scenario: "I insist that what scientists share is not sufficient to command uniform assent about such matters as the choice between competing theories" (Kuhn 1962/2012, 185).

Here's the thing, though: *uniformity happens.* It's what happens when some "achievement was sufficiently unprecedented to attract an enduring group of adherents away from competing modes of scientific activity." It's what happens when "scientific communities . . . reach a firm consensus unattainable in other fields"; and when "consensus endure[s] across one paradigm change after another" (Kuhn 1962/2012, 172). It is an empirical fact that entire groups of scientists occasionally transition en masse to the use of a new paradigm/theory. Thus, even if we accept Kuhn's argument that consensus on what constitutes "good reasons" for paradigm choice cannot logically mandate a specific choice across members of the scientific community, all this would show is that

shared values cannot by themselves explain the uniformity those ·
members exhibit. It would not show that there is no uniform motion of members. But, unless we think it is a coincidence that keeps
happening over and over again, *something* must explain the uniformity. To be fair, Kuhn never suggested otherwise. Uniformity
is a cornerstone of the Kuhnian vision. His point about the heterogenous application of shared criteria for choice is, I think, not
designed to undermine the empirical fact that uniformity happens,
but rather the presumption that the uniform motion of members is
a consequence of a rational decision logically imposed on members
by virtue of their embrace of certain criteria for choice. In other
words, even if "what scientists share is not sufficient to command
uniform *assent*," it might still be sufficient to command something
more cognitively passive than rational assent. And, indeed, Kuhn
confesses to the surprise of no one at the very end of "Objectivity,
Value Judgement, and Theory Choice" that "an individual's transfer
of allegiance from theory to theory is often better described as conversion than as choice" (338). One of *Structure*'s main points, of
course, was that new paradigms/theories are not really *chosen* at all.
And yet, shift happens.

To my enduring frustration, Kuhn never floated an alternative
explanation for uniformity. The Laudans (1989) did, though. They
argued that there's simply no tension between the empirical fact of
uniformity and the diversity of values. You see, what happens is that
scientific communities achieve consensus when a theory satisfies
the full range of the membership's choice criteria. Early adopters,
for example, aren't overly insistent on the accumulation of predictive successes; later adopters are (or might be). In any event, once a
theory possesses all the virtues it needs to impress everybody, consensus is achieved. Case closed. The muscle of this argument is its
elegance; it turns the philosophical irritation created by human diversity into a strength of sorts. Rather than allowing subjectivity to
threaten the rationality of science, it reveals theories that achieve
consensus to be *normatively robust*, to coin a term that is definitely

going to catch on. We should have *extra* confidence in theories that achieve consensus, because only a theory that was stupefyingly virtuous could pull off such a feat. Obviously Kuhn would have hated this argument, because it retains the element of choice that he maintained was not quite faithful to how paradigms are acquired. But who cares? If he had hoped to use the diversity of values as a bludgeon with which to cajole us into accepting the noncognitive nature of theory choice, well, the Laudans put the kibosh on that.

The problem is that the Laudans' model cannot accommodate the fact that things appear not to develop historically in the way their argument requires. To take one example, Kuhn and the Laudans agree that the investigation of novel predictions made by a theory can only occur at relatively advanced stages of theoretical development. The Laudans use this plausible claim to argue that practitioners who value novel predictions in particular when deciding whether to swear allegiance to a theory can be expected to embrace it at a relatively later period compared with other, alternatively inclined members of the community. This is the crux of the disagreement. In *Structure*, the investigation of novel predictions is something that takes place during a period of *normal science*—that is, after some achievement has attracted an enduring group of adherents away from competing modes of scientific activity. Once a group reaches that point in Kuhn's model, the transition is essentially over. At that point, practitioners are joining an established research tradition. The epistemic role of investigating novel predictions under these auspices, as Kuhn argues in chapter 3 in *Structure* and in his contemporaneous "The Function of Measurement in Physical Science," is not so that practitioners can finally make up their minds about whether to work within that tradition. Those kinds of investigations are only possible *because* the decision period is over and because there is now a tradition within which to work and within which to train incoming generations of scientists.

Characteristically of *Structure*, this claim is simultaneously radical and, upon reflection, hard to deny. But it is deeply unintuitive.

Why *are* scientists investigating novel predictions, if not to test theories in order to see whether they are true, and thus command our rational assent? Kuhn has some interesting things to say about this, which we'll look at later. But we shouldn't let his answers bear on whether we're convinced that paradigm/theory allegiance precedes the investigation of novel predictions. We have much higher quality evidence to which we can appeal.

The first piece of evidence comes in the form of Stephen Brush's (2015) systematic study of whether a theory's ability to make novel successful predictions matters more to its adoption than its ability to accommodate established relevant facts. The specific question to which Brush sought an answer was whether novel successful predictions made a difference to the reception of a theory. His approach was to look at the time at which the theory achieved consensus. If it was before any predictive successes were recorded, then the community's consensus did not depend on predictive success. Out of more than a dozen instances of theories that were known to have achieved predictive successes, he found only one case in which consensus followed predictive success: Mendelev's periodic law. (I will argue later that even this case is highly equivocal.) For the rest, the predictive successes followed consensus, sometimes by several decades. By contrast, the time intervals required for the community to come together were short, which agrees with Kuhn's depiction of the route to normal science in "The Route to Normal Science." For instance, Brush argues that any meaningful debate over general relativity had ended before the 1920 publication of the 1919 Eddington expeditions to Príncipe and Brazil. Our lore and our epistemological biases insist that Eddington's solar eclipse observations *had* to matter to the community's allegiance, and that, in general, the absence of successful predictions should impede consensus. But the historical facts suggest otherwise.

In particular, what history suggests is that these transitions are much quicker than we'd expect them to be if the Laudans were right—that is, if consensus doesn't occur until everyone's rational choice criteria are satisfied. Not only does uniformity happen, but it characteristically happens on a timescale and upon an evidence base that makes no sense if we assume that the way practitioners develop an allegiance to a paradigm/theory is by way of a process that looks like rational choice guided by the canonical epistemic norms. It actually looks much more like the cognitively passive process Kuhn originally envisioned. We know that the moments of "choice" that preoccupied Kuhn's critics were not exactly emblematic of the process of paradigm acquisition in general as he imagined it. Rather, that process is exemplified by scientific training. There's a good reason for that: almost all instances of paradigm acquisition occur through learning. The revolutionary "choice" points are few and far between. In the pedagogical context, very little of what is acquired can be described as having been chosen. It's simply learned. Physicists in training do not *adopt* Newtonian mechanics, nor do they become *convinced* of it. But they acquire it nonetheless. That process starts something like this: someone draws a triangle on a chalkboard. She spends a little while getting students to see the length of a triangle's arm as representing the magnitude of a velocity, and its spatial orientation as representing a direction of motion. Then she shows how to set up a trigonometry problem that will allow them to calculate other magnitudes and directions. Then she writes out an equation and walks them back and forth between the vector diagram and the equation's symbols. She goes through a couple of examples. And . . . they're off.

Reflect for a moment on what has *not* happened. Students have not been asked whether they find Newtonian mechanics convincing, or why they should believe that forces can combine vectorially (no one actually knows; see Lange 2011). They haven't been presented with any reason for believing anything about nature. No attempt has been made to persuade them of the truth of Newton's laws. Sometimes you get a cute little demonstration of this or that. But is the point of this to win over any remaining holdouts, the few rebels who aren't just going to take her word for it? Of course not.

Because the students are not there to be convinced or to buttress any beliefs. They are there to learn how to solve the kind of physics problems on which they're going to be tested. The form of instruction does not lend itself to critical thought or assessment, nor is it designed to. It is designed to impart a conceptual and mathematical framework for representing certain kinds of phenomena. And in this it has succeeded. The further they go through their program, the further entrenched they become in the game of representing, formalizing, and calculating. Pointless questions about truth and reality quietly fade into oblivion. All is well. Paradigm acquired.

In the general case, this is how paradigm allegiance happens; the confrontational "theory choice" scenario preferred by philosophers of science is highly nonstandard. And, unlike that nonstandard scenario, the pedagogical transmission of paradigms is manifestly effective at producing the characteristic uniformity that prevails across a discipline. Most importantly, it is manifestly capable of producing that uniformity in the absence of appeals to the canonical epistemic values. It has to be—after all, it is through this training process that one begins to develop an appreciation for properties like accuracy, simplicity, and so on. If we agree that the majority of productive and creative scientific work emerges out of this context, why should we be scandalized by the notion that there might be instances in which the rational choice criteria do not exert control over the direction of inquiry? From the perspective of the normal paradigm acquisition process, it is not difficult to see, for example, how utterly meaningless a successful prediction might be for someone who is bred to approach physical "theory" as a set of techniques for solving physics problems. Even if we grant that a successful prediction bears logically on a theory's truth, the physicist trained in the tradition of that theory is probably not accustomed to thinking of the theory in terms of its truth or falsehood. If she rather thinks of it as the style or framework in which she works, the logical connections between it and experience needn't remain constantly at the front of her mind. What concerns her is whether she can and should continue to work in that framework, and the theory's truth is not necessarily relevant to settling those issues. Philosophers have always wanted theory choice to be a choice regarding what to believe about nature. Kuhn believed that scientists saw it as a choice about what to work on, about "which paradigm should in the future guide research." These are not equivalent. Even if a theory's successful prediction implies its truth, the scientist is not required to care.

2.4. The Significance of Setup

For the scientist, the question of "which paradigm should in the future guide research" is not first and foremost a question about which theory is most likely to be correct. It is a question about which paradigm "promises the concrete successes for which scientists are ordinarily rewarded." Kuhn understood a concrete success to be a "solved problem," which he called "the unit of scientific achievement" (1962/2012, 168). In *Structure*, he is singularly focused on the notion that a challenger paradigm's concrete success in resolving an empirical anomaly is the primary cause of the community's transition to it; it is what grounds the promise of further concrete successes. But the book offers no explanation for how this could possibly work. Indeed, by invoking faith, he hoped to indicate that there was no rational explanation.

While the "faith" debacle was the only time where Kuhn appears to have been cowed into submission, it is but one of several instances in "Postscript" where he has clearly rethought the direction of a prominent path taken in *Structure*. Kuhn's commitment to the inadequacy of the canonical values for determining the rational choice of paradigm/theory was ultimately rooted in the fact that he wanted moments of paradigm *transition* to be a species of the same genus to which trainee paradigm acquisition belongs. The essence of that genus is distinguished by two principal features.

One is the development of a perspective on how inquiry ought to be conducted. In the trainee context, this process of development is clearly not guided by articulated norms of rational choice. It is guided by exemplars. The other is the problem-solving capacities enabled by that perspective. In the trainee context, problem-solving is aided not by true beliefs about nature, but by *the ability to set up problems*. Each of these dimensions saw renewed attention in "Postscript."

As Kuhn sheepishly observed, *paradigm* is a rather badly be-haved term in *Structure*, at least by the standards of mid-twentieth-century analytic philosophy. He took the hectoring he received with good humor. He also took seriously the task of providing a refined and more circumscribed version of whatever it is that "define[s] the legitimate problems and methods of a research field for succeeding generations of practitioners" (Kuhn 1962, 10). The capacity of a paradigm to function as the arbiter of legitimacy in science was for Kuhn its single most important feature; this, more than anything, is what gives rise to what he called *normal science*. It makes sense that, when faced with demands for clarification, he chose to focus his efforts here. In the end, the version of "paradigm" which best captured this dimension was that of a "shared example" or *exemplar* (Kuhn 1962/2012, 186–190; see also Kuhn 1974; Hacking 2016).

As a way of articulating this more focused conception, Kuhn returned to a brief discussion in *Structure* regarding the status of $f = ma$ among physicists, which had interested him there because of the way in which it "behaves for those committed to Newton's theory very much like a purely logical statement that no amount of observation could refute" (Kuhn 1962/2012, 78). He felt this supported his claim that anomaly provides insufficient inducement to reject a paradigm. Kuhn was drawing on an extended discus-sion of the multifaceted life of $f = ma$ that appeared in Norwood Hanson's *Patterns of Discovery* some years earlier. And, indeed, Hanson does describe the status of $f = ma$ as working more as a definition of 'f' than an empirical claim that could be subject to experimental refutation. However, this specific, definitional feature of its behavior is not center stage in Hanson's account; much of his analysis is focused on the role of $f = ma$ as a schema for organizing specific kinds of information in order to facilitate analysis and computation in mechanics (Hanson 1958, 99–105).

By the time he wrote "Postscript," Kuhn had clearly reinterpreted the significance of $f = ma$ from the perspective of this other theme running through Hanson's discussion. While his remarks in *Structure* do not rule out this alternative function of the law, they are concerned only with the psychology of theory and evidence in scientific practice—or, more properly, with undermining a certain widespread view of how disconfirming evidence ought to be treated when science is done correctly. The definitional status of $f = ma$ is salient in this context. But as an engine for facilitating the emergence of normal science—which is what a paradigm is supposed to do—the fact that some set of symbols is treated functionally as a definition is neither here nor there. Rather, what seems to allow $f = ma$ to play its role as an essential component of normal physical science is the way in which it is treated as "a law-sketch or a law-schema":

> As the student or the practicing scientist moves from one problem situation to the next, the symbolic generalization to which such manipulations apply changes. For the case of free fall, $f = ma$ becomes $mg = m(d^2s/dt^2)$; for the simple pendulum it is transformed to $mg \sin \theta = -ml(d^2\theta/dt^2)$; for a pair of interacting harmonic oscillators it becomes two equations, the first of which may be written $m_1(d^2s/dt^2) + k_1s_1 = k_2(s_2 - s_1 + d)$; and for more complex situations, such as the gyroscope, it takes still other forms, the family resemblance of which to $f = ma$ is still harder to discover. (Kuhn 1962/2012, 188)

What we find in physics, in other words, is that a certain way of relating highly general but physically significant categories can take a variety of specific forms depending on the kind of physical system one happens to be dealing with. Sometimes we use it to symbolize the relationship of a pendulum's length to its period. Other times we use it to represent the distanced travelled by a falling object in relation to the time it's been falling. It combines accuracy and elasticity in a way that makes it possible for the practitioner to do several jobs with a single tool, if she so chooses.

Viewed from this perspective, it is easy to understand why Newton's second law would have appeared to function "as a purely logical statement that no amount of observation could refute." In practice, it is not really a candidate for refutation, because practitioners do not psychologically relate to it as a statement, logical or otherwise. It is more like a pattern or template to which one tries to conform when setting up a problem in mechanics. Nested within the pattern are references to categories represented by the symbols 'f,' 'm,' 'a,' and '$=$.' But the referents of those symbols are not fixed in every context. In those contexts in which the referents *are* fixed, they are fixed by prior community practice; a practitioner's scientific duty in these contexts is to conform to that tradition. In those contexts in which the referents are *not* fixed, the practitioner's goal is to find creative yet useful ways of conforming to the pattern—that is, to find new referents of 'f,' 'm,' and 'a' (probably not '$=$') that satisfy the relevant relations in a way that captures some phenomenon that the community recognizes to be important. In each context, the practitioner treats $f = ma$ as an exemplar upon which he models his own problem-solving efforts. As we will see, these two types of context taken together summarize a great deal concerning how a paradigm exercises its power to confer legitimacy on a given scientific activity. *Something* about it—Kuhn never says what—imbues it with a normative significance. It doesn't just tell practitioners how to relate this to that and so forth. It tells them those categories are of particular importance for understanding physical phenomena.

This brings us to another dimension of scientific inquiry about which Kuhn had substantially refined his thinking leading up to "Postscript." The confluence of Newton's second law functioning as an exemplar and its facilitation of setting up problems is no accident. The centrality of problem setup to the development of scientific knowledge does not receive any explicit attention in *Structure*. It lurks in the murky depths of chapter 2, "The Route to Normal Science," but it never really breaches the surface. Because the solved problem is what scientists are rewarded for, and because his model is designed to treat with the utmost seriousness the psychological reality of the practicing scientist, Kuhn is understandably more focused on the influence of solved problems; the solved problem is, in his words, "the unit of scientific achievement." The tacit conceit of *Structure* is that an "achievement . . . sufficiently unprecedented to attract an enduring group of adherents away from competing modes of scientific activity"—one of two defining characteristics of a paradigm—is a solved problem, or "a few" solved problems (Kuhn 1962/2012, 10).

"Postscript" reflects a stark recalibration in Kuhn's views concerning the factors governing paradigm choice. *Structure*'s silence on setup has given way to the uncharacteristically forthright claim that, "at times of theory-choice . . . *the demonstrated ability to set up* and to solve puzzles presented by nature is, in case of value conflict, the *dominant criterion* for most members of a scientific group" (Kuhn 1962/2012, 204; emphasis added). This comment follows an extended discussion of the process by which students learn how to solve textbook problems, along with a glance back at the circuitous legacy of Galileo's solution to the problem of pendular motion. In both instances, Kuhn's emphasis is on the supreme significance of the practitioner's ability to provide a problem with a kind of structure, to set it up so as to make the basic form of the solution perceivable and the solution itself imminent. Of related interest in this context is how the habitual practice of setting up problems in a certain way affects the practitioner's perception of nature. Physicists adept at setting up problems on the pattern of $f = ma$ begin to literally see nature in terms of forces, masses, accelerations, and the way those properties relate to one another. This enables them, on Kuhn's view, to extend the reach of $f = ma$ to novel scenarios. All of this receives more in-depth treatment in his contemporaneous "Second Thoughts on Paradigms," which "Postscript" describes as having been published in "1970 or 1971" (actually 1974).

In this way, problem setup functions as something like a keystone for Kuhn's entire vision of science. On the one hand, it is the primary thing that exemplary works exemplify. On the other hand, it eventually warps practitioners' experience of the world in the way Kuhn requires for the detection of anomaly and the consequent precipitation of crisis. According to his way of thinking, in attempting to analyze a novel phenomenon by shoehorning the analysis into the setup that is characteristic of pendular motion, I (or my successors) will eventually *come to see* that novel phenomenon as a pendulum.

More importantly, though, the idea that exemplars exemplify how to set up problems fills the key lacuna in Kuhn's account of paradigm choice—you know, the lacuna he hoped to leave open in order to preserve the arationality of those choices. Whereas *Structure* cannot quite explain how a solved problem results in the perception of promise, what we can now see is that it isn't the solved problem per se that makes the challenger paradigm look so promising. It's the setup.

One of the principal trials Kuhn faced as an intellectual was that he was an epistemological naturalist way before it was cool. The all-important role afforded to the actual psychology of the individual practitioner in *Structure* was/is unpalatable to philosophers of science. For him, though, there could be no other way. The "pattern of scientific development," as he called it, is not generated by Bayes' theorem, nor by *modus tollens*. It is produced by the actual decisions of actual scientists, who may or may not recognize the significance of such things at any given moment during the course of inquiry. On his view, the project of understanding the episte-mology of science is one of constructing a causal model that takes as its input the psychology of a practicing scientist and generates as its output the gross pattern of the history of science.

The scientist in Kuhn's story is driven fundamentally by the question of what to work on; it is what "attracts an enduring group of adherents away from competing modes of scientific activity." Perhaps it is here that we see just how different the Kuhnian vi-sion was from dominant modes of philosophy of science. The way in which this question is resolved strongly affects the devel-opmental trajectory of scientific knowledge. As such, it must fea-ture centrally in a Kuhnian epistemology of science—even if that means diluting the rational purity of scientific practice. In contrast, few philosophers of science working today would be comfortable with the idea that something as mundane and idiosyncratic as one's choice of research problem is philosophically relevant, let alone a matter of grave epistemological significance.[2] Such considerations, it is thought, belong to the untidy world of idiosyncratic prefer-ence, driven by epistemologically indecent motivations like per-sonal glory or native curiosity. Whether that's true or not, the fact is that these factors are highly influential parts of the causal process that results in what we call "scientific knowledge." Nothing is going to change that. We can decide that the way science actually works is irrelevant to our philosophical ambitions, that what we as philosophers of science are interested in is crafting a more perfect version of scientific inquiry. Maybe that's admirable and advisable. But it often seems that philosophers of science do not want to im-prove scientific inquiry. They want to *replace* it with something which, like "the methodological stereotype of falsification," bears little resemblance to anything "yet disclosed by the historical study

---

[2] The pioneering efforts of Kitcher (1993, 2001) and subsequent expansion and refinements of Zollman (2010) and Weisberg and Muldoon (2009) are welcome exceptions.

of scientific development" (Kuhn 1962/2012, 77). There is plenty that one might say about the philosophical or otherwise intellectual coherence of such a position. But no one ever listens, so why bother? Bayesians gonna Bayes . . .

Whether one ought to work on something is constrained by whether she thinks she can produce something that satisfies the community's criteria for success—that is, whether she thinks she can produce a solved problem. What Kuhn came to appreciate after *Structure* was that information about fruitfulness, about "the promise of concrete success," cannot be gleaned solely from the fact that an exemplary problem has been solved. A scientist's choice of what to work on, like her reasoning in the context of scientific training, is dominated by considerations of whether she has a means to set up the problem. Many perfectly coherent questions about nature, even "well-posed problems," fail to satisfy this criterion. This suggests that a paradigm's power inheres not in the fact that it solved a problem, but in the fact that it facilitated the setup of a problem in such a way that it *could* be solved. Her belief in the promise of a paradigm will be determined by the degree to which it appears to offer the opportunity for producing solved problems. Of all the factors that affect the choice of what to work on, Kuhn singled out this promise as being "of special importance in actual scientific decisions" (Kuhn 1977, 322, fn.6). In "Objectivity, Value Judgment, and Theory Choice," he chose the term *fruitfulness* as a label for "the promise of concrete success."

2.5. The Puzzle Refined

I began this chapter with the claim that the Puzzle's essence is the search for an alternative species of epistemic warrant that satisfies a variety of constraints imposed upon us by the history of science. We've surveyed the evolutionary history of these constraints, and

I now want to review them in close proximity to one another in hopes of bringing the Puzzle into sharp relief.

1. The first and most general constraint is that the species of epistemic warrant in question must hold, as Kuhn says, "in defiance of the evidence provided by problem-solving" (Kuhn 1962, 156–157). This is what makes the Puzzle philosophically interesting; it ties our epistemological hands in an unusually strong way, barring any appeal to a track record of success. It is also, I believe, what makes the Puzzle realistic. Much of what I do in this book involves simply describing instances from the history of science in which choices were, in fact, made in this way. The philosophical challenge is to articulate a conception of epistemic warrant that can validate these choices as epistemically justified or well-founded.

2. Information about what to work on must be extracted from an exemplar.

3. The basis for decisions about what to work on must be generally accessible to the scientific community at a very early stage of the development of a paradigm/theory. This is the constraint imposed by two historical facts: (1) that shift happens, and (2) that shift very frequently happens *en masse*. The timescale for many transitional episodes is stunningly short.

4. Epistemic values like accuracy, simplicity, and so on play a significant role in decisions about what to work on. But they are not the whole story. There is an as-yet unarticulated factor that motivates these decisions. (Yes, I'm going to articulate it. Later.)

5. Whatever does settle the matter must be capable of causing the growth of scientific knowledge. Because of scientific investigation, we know more about the natural world than we used to. Why? What is it about scientific investigation that causes this to happen? In particular, what is it about the way in which scientists decide "how to go on" that causes it?

How is it that scientists, in Wittgenstein's words, "know how to go on"? Kuhn thought they didn't know—that they *couldn't* know. I join the majority of philosophers of science of the last six decades in rejecting this Kuhnian misstep. They definitely know *something* about how to go on. The history of modern science testifies to this. But that same history also testifies to the fact that what they know about how to go on is not of the kind of knowledge we are accustomed to associating with scientific inquiry. It is not supported in the traditional way—or, at least, it is not supported in the way we traditionally imagine scientific knowledge to be supported.

To understand how it is that scientists know how to go on, we must understand the nature of fruitfulness. A fruitful theory does not just solve problems. It does not even primarily do that. It "leaves all sorts of problems for the redefined group of practitioners to resolve" (Kuhn 1962/2012, 11). Readers of Newton's *Principia*, or Franklin's *Electricity*, or Darwin's *Origin*, did not just see solved problems. They did not even primarily see solved problems. They saw a veritable engine for *generating* problems that they themselves could solve. They saw its future promise. A fruitful paradigm/theory has a certain propensity to generate solvable problems. The solution to the Puzzle of Promise lies in understanding what gives a paradigm/theory that propensity, and how practitioners are able to reliably assess a theory's possession of it. Fruitfulness, just as Kuhn said, "deserves more emphasis than it has yet received" (Kuhn 1977, 322, fn.6).

3

The Perception of Promise

The [scientific] enterprise cannot be thoroughly unpredictable. Otherwise it would be impossible to proceed rationally or productively. If there is "no way to tell where a given domain of research will lead," investigators must, nevertheless, have reasonable confidence that their pathways will lead them to something significant enough to reward the great investment of effort they must devote to their work. Otherwise the scientific enterprise could not have flourished as it has done for the past three centuries.

—F. L. Holmes, *Investigative Pathways*, 143

3.1. The Nexus of Normativity

The triumph of *Structure*'s early chapters is to draw attention to the normative status and powers that works of science sometimes acquire:

> Aristotle's *Physica*, Ptolemy's *Almagest*, Newton's *Principia* and *Opticks*, Franklin's *Electricity*, Lavoisier's *Chemistry*, and Lyell's *Geology*—these and many other works served for a time implicitly to define the legitimate problems and methods of a research field for succeeding generations of practitioners. (1962/2012, 10)

Now that's a very interesting thing to be able to do, isn't it? Through a traditional epistemological lens, we can understand readily enough

Fruitfulness. Chris Haufe, Oxford University Press. © Oxford University Press 2024.
DOI: 10.1093/oso/9780197666395.003.0003

why a research method might come to be regarded as legitimate—it's reliable, or truth-conducive, or some other standard notion. But the legitimation of a research *problem* is not so easily captured. Traditional epistemological categories are designed for the appraisal of answers, not questions. Nor does the power to *implicitly* define legitimacy fit particularly well with familiar approaches to knowledge. For the historian, though, the implicit legitimation of research problems is the most familiar thing in the world. It is also among the most consequential for the growth of knowledge. No one who has studied the history of science, particularly in the modern period, can doubt the awesome power of certain achievements to constrain the path of inquiry. The historian could no more disavow the epistemological significance of this phenomenon than she could that of research methods. Kuhn was among the first to appreciate the significance of this phenomenon, and to trace its all-encompassing effects.

The epistemological radicalism of *Structure* lies in the fact that legitimacies of methods *and* of problems derive from one and the same source. This makes it difficult to retain our traditional answers to what makes a method legitimate, because a research problem is not legitimized by anything like its reliability, whatever that could mean. But it is legitimized by *something*. For, every capable scientist knows a good research problem when she sees one, and scientific communities exhibit broad agreement on which are the good problems. They also discriminate harshly against those problems that aren't so good. So much of *Structure* is fixated just on establishing the reality of considerable normative constraints on inquiry as a fact of scientific life, and on the way in which textbooks and works of science function as the ultimate source of those normative constraints. Almost nothing is said about how these normative constraints emerge, however. In his most specific mood, Kuhn says that the way some works of science acquire their normative powers is on account of two essential characteristics. Their achievement was sufficiently unprecedented to attract an enduring group of adherents away from competing modes of scientific activity. Simultaneously, it was sufficiently open-ended to leave all sorts of problems for the redefined group of practitioners to resolve. (Kuhn 1962/ 2012, 10–11)

That's a bit brief. Of all the things left unsaid here, though, the one that has gnawed at me incessantly is the matter of *how* a work of science flexes its irresistible powers of attraction. What are practitioners *getting* from these works such that they abandon the life they've built in order to start a new life across town with a much younger paradigm? Nickles (2013, 112) charges Kuhn with being "unable to explain in any detail" how this occurs, owing to some intrinsic features of his account. Hacking (2016), rather more generously, forgives Kuhn for failing to supply the answer to a question on which Aristotle also punted.

Not to be too dramatic, but any hope of articulating the alternative species of epistemic warrant we seek hinges on being able to offer an account of this process. Whether or not Kuhn had in his copious mind a concrete image of what exactly practitioners acquired from these great works of science, or how they acquired it, he never shared it with the world. Perhaps his actual view was as irredeemably gappy as Nickles describes in his claim that it amounts to nothing more than "a sudden 'aha' experience" (Nickles 2013, 112). And perhaps this explains in part the attraction that the concept of faith held for him in these contexts. What is beyond dispute is that the details of the process are not specified in *Structure*. In the previous chapter, I argued that Kuhn seemed to recognize this, focusing in later work on exemplars and the set-up of problems. This was progress. However, nowhere did he offer an account of how exemplars emerge, nor of how knowledge of set-up was acquired. Nor did he ever deviate from his conviction that a solved problem is the flashpoint for the explosion of normal scientific activity. The search for an alternative species of epistemic warrant begins with the rectification of these defects. We need to understand where the normative pressure in science comes from, and what it is that practitioners strive to respond to.

In this chapter, I introduce the engine that drives my account of fruitfulness: the *family of solvable problems*. Before any change in worldview could or did occur, the great works of science effected the emergence of a new kind of scientific problem, characterized by a certain set-up or structure. The imposition of structure does not solve problems; it makes them *solvable*. The fine-grained details of these structures are subject to various types of manipulation, a fact that practitioners routinely exploit in their explorations of nature. This tendency to exploit the ambiguities of structure results in groups of related problems that I call *families*. The notion of a family of solvable problems sits at the juncture where promise is perceived, normative constraints emerge, and the causal powers that promote the growth of knowledge are unleashed. Whereas traditional conceptions of epistemic warrant take the goal of scientific inference to be (approximate) truths, the alternative species of epistemic warrant we seek has as its object the adoption of a perspective that results in a family of solvable problems.

I first want to lay out the basic historical process through which normative constraints emerge from works of science, focusing on the details of how a solved problem might support the perception of fruitfulness potential. It will turn out that the focus on solved problems as exemplars is unmotivated, because exemplars also emerge from *failed* problem-solving attempts. The specific ways in which the "solved problem" version of the process fails reveal new details about the emergence of exemplars and, by extension, of normative constraints on inquiry. In particular, it provides strong support—stronger than Kuhn ever provided—for the notion that paradigm/theory choice is not about problem-solving success. More importantly for my purposes, it provides strong support for the idea that the potential for fruitfulness can be judged prospectively, "in defiance of the evidence provided by problem-solving." It's what Kuhn would have wanted.

3.2. Problems Solved

The opening salvo of *Structure* is the idea that those classic works of science mentioned earlier solved some important problems and, in doing so, grabbed the attention of practitioners working on related topics. It is a matter somewhat open to interpretation, but it is my view that Kuhn uses the notion of "sufficiently unprecedented achievement" to refer to solved problems. For one thing, he claims that "the unit of scientific achievement" is the solved problem. Second, he mentions repeatedly that challenger paradigms are chosen after (and only after) having solved a few problems. Third, he cites "concrete puzzle-solutions" as the "basis for the solution of the remaining puzzles of normal science" (Kuhn 1962/2012, 174). Fourth, his illustrations of something serving as the basis for such solutions involve solved problems (e.g., his discussion of Galileo's study of the inclined plane [see later], or the discovery of Uranus). Fifth, he never hints at any other route by which a work of science might come to "attract an enduring group of adherents away from competing modes of scientific activity." In Kuhn's view, these works become exemplars after and because they solved problems.

For the purposes of this discussion, let us say that a problem is taken to be solved when *the relevant community of practitioners comes to embrace a particular solution to that problem as satisfactory enough to warrant the cessation of further scientific debate.* Let us further label this state *consensus.* I know this definition will be too psychological or sociological for some people. But, for better or for worse, science is practiced by humans; some such element or other must come into the picture if we are to explain how a group of humans behaves. And in any case, we have to proceed down this path if we are trying to reconstruct Kuhn's implicit theory of exemplar emergence, owing to the fact that his image of scientific development is a causal model driven partly by the psychology of human perception and decision-making, and partly by the socio-cultural factors that promote and maintain normative constraints on modes of scientific investigation—that is, on which methods and problems are regarded as legitimate. The specific epistemic conditions that need to be met for the community to achieve consensus on the solution to a problem are not uniform; what counts as "satisfactory enough" to be getting on with will vary across practitioners and across communities. But the *effects* of consensus exhibit a pleasing and important generality. In the sciences, a solution around which consensus has been reached functions as a kind of constraint on future inquiry, whose increasing breadth and depth require an ever-expanding foundation from which to launch promising new lines of investigation. Scientists are rewarded for solving problems not least because of the critical function that solved problems perform in the ongoing growth of scientific knowledge.

The solved problem is a seductively intuitive starting point for thinking about how exemplars emerge. Newton's *Principia* was understood to solve the pressing seventeenth-century problem of what form a force would have to take to produce an elliptical orbit. It was understood to solve—at some level, anyway—the pressing problem of the sun's causal contribution to orbital motion. And many other problems besides, both in mechanics and in mathematics. To say that this achievement was "unprecedented" is to understate things to the point of folly. It is, quite simply, astounding. Consider the French mathematician L'Hôpital's totally appropriate reaction upon being shown Newton's solution to an outstanding problem in the mathematics of fluid resistance:

> he cried out with admiration Good god what a fund of knowl-
> edge there is in that book? He then asked the Dr. Every particular
> about Sr I. Even to the color of his hair said does he eat & drink
> & sleep. is he like other men? & Was surprised when the Dr told

him he conversed chearfully with his friends assume nothing &
put himself upon a level with all mankind.[1]

Unsurprisingly, practitioners in these and adjacent fields were,
within a generation of its publication, working in the tradition defined by the *Principia*.

Now, what is it about solving important problems that turns a
work like the *Principia* into an exemplar upon which they pin their
hopes for future inquiry? Kuhn does not say what it is in *Structure*.
Given his understanding of how crises are resolved, Kuhn probably
thought that it was just obvious that a theory's demonstrated ability
to solve one or more important outstanding problems would leave
the crisis-worn community with no choice but to adopt it, perhaps
encouraged by the vague principle that nothing succeeds like success. The desperation that Kuhn attributes to practitioners in a crisis
state makes palpable the sense of relief they must experience when
such a problem is solved, and against that backdrop it does indeed
seem obvious that solving the problem(s) is sufficient inducement
to climb aboard. Who in their right mind (among the English, anyway) would not have adopted the Newtonian framework after
nearly a century's worth of anxiety over why planets travel in
ellipses around the sun? The promise of the Newtonian paradigm
inheres in the fact that it succeeded where nothing else had. Similar
stories might be told for Darwin's *Origin*, Lavoisier's *Chemistry*,
Franklin's *Electricity*, Lyell's *Geology*, or Ptolemy's *Almagest*, each
of which purports to solve a number of empirical problems that
had dogged specific communities for extended periods of time—in
some cases several centuries (Ptolemy), even millennia (Darwin).

But it's not actually all that obvious that this should work. What
would it mean, for example, to say that Darwin's *Origin* solved the
problem of adaptation? Per our earlier definition, it would mean

[1] Keynes MS. 130.05, King's College, Cambridge, UK, www.newtonproject.ox.ac.uk/
view/texts/diplomatic/THEM00168. Quoted in Westfall (1980), 473.

that the community of naturalists determined that descent with modification through natural selection was satisfactory enough as an explanation for how adaptations arise that there was no use in arguing about it anymore.[2] So, now what? If this is the point at which promise is perceived, there must be something about the mere fact that Darwin's theory has solved the problem that can, all by itself, provide some inkling as to what lies in store for future inquiry. Here's a suggestion that Kuhn probably would have hated: solving the problem of adaptation constrains future inquiry by adding to the store of empirical knowledge to which future theories are responsible. Here we imagine scientific knowledge to be a set whose elements must be logically consistent with one another. Solved problems become elements of this set, and any future candidate elements are constrained by the demand for consistency with that set. The vestigial logical positivist that resides in most of us will be strongly attracted to this sort of picture. But it cannot be an important part of the story of how these achievements actually constrain future inquiry, because logical consistency is simply too weak of a normative constraint to explain the aspects of scientific practice that are responsive to the achievements invoked by Kuhn. Their defining feature is their capacity to *generate* new lines of inquiry. For all its virtues, logical consistency cannot tell us what we ought to work on, nor how we ought to work on it. Mere logical consistency leaves too many investigative paths open, and therefore cannot account for the overwhelming salience of particular pathways subsequent to a problem's being solved. To explain the cognitive role of solved problems, we must give some account of their constraining power that allows them to exert far greater influence over the process of inquiry. I will argue that fruitfulness considerations are primarily about the search for a species of

---

[2] Gayon (1998) marked this point somewhere in the early 1930s; Brush (2009) sets it in the mid-1940s.

constraint that is far more powerful, wide-reaching, and explanatory than anything to which logical consistency could aspire.

3.3. Problems Classified

The failure of logical consistency to explain the effect of certain
works of science on subsequent inquiry sheds some much needed
light on the dark space where promise is perceived. Kuhn attributes
to these exemplary works the defining characteristic of being "sufficiently open-ended to leave all sorts of problems for the redefined
group of practitioners to resolve." But this is easier said than illustrated. His illustrations tend to focus rather narrowly on the further
development of specific lines of inquiry, such as the increasingly
precise measurement of physical constants, the development of apparatus like the Atwood machine (a system of masses and a pulley
that can be used to study constant acceleration), and the "articulation" of theoretical and observational consequences of the new
paradigm. This is Kuhn's famous picture of normal science, which

> consists in the actualization of that promise [of success discover-
> able in selected and still incomplete examples], an actualization
> achieved by extending the knowledge of those facts that the par-
> adigm displays as particularly revealing, by increasing the extent
> of the match between those facts and the paradigm's predictions,
> and by further articulation of the paradigm itself. (Kuhn 1962/
> 2012, 24)

The Kuhn who wrote *Structure* tended to think of normal science
as essentially a dynasty of follow-up questions that just sort of
naturally arise once the community adopts a paradigm, the "further problems" left for "the united group to resolve" (Kuhn 1962/
2012, 23).

In what way are these follow-up questions the "actualization of a promise of success" that inheres in exemplary works of science? I grant that such follow-up questions do naturally arise; for many practitioners, there will be an instinctive logical progression from accepting that there is a gravitational constant to attempting to determine its magnitude. But I have a hard time accepting the idea that seventeenth-century readers of the *Principia*, or nineteenth-century readers of the *Origin*, were instantly bewitched by the specific follow-up questions that they were now licensed to ask. Rather, something like L'Hôpital's situation seems to be more representative: there is an antecedent interest in some phenomenon, and the exemplary works of science successfully tackled the phenomenon in a novel way. Those works served as a bridge that connected the practitioner to the promised land from which she has long been separated by a gaping chasm. That hitherto unreachable destination is already populated with loads and loads of problems to work on, even well-posed problems. It is not as if she has been lolling around, idly wishing that there was something about nature that escaped her grasp so that she might have something to investigate. The capacity for follow-up questions is not limited to those epoch-defining works of science on which Kuhn's story relies. Insofar as these lines of inquiry arise out of the logical implications of the relevant paradigm, there is nothing in this respect that differentiates the exemplary works from the "competing modes of scientific activity" away from which an enduring group of adherents has been attracted. They too were pregnant with the kinds of theoretical and empirical questions that inevitably arise once one has elected to go some distance down a reasonably well-defined path. Anyone who wants to see what the tightly governed elaboration of theory looks like need look no further than the early modern Aristotelians who "mopped up" a broad range of specific questions left by their medieval predecessors (Joy 2006, 82–87). So the promise of success cannot simply inhere in the promise of questions open for investigation. As every parent knows, no statement is immune from follow-up questions. It was not for lack of follow-up questions that, for instance, Steven Weinberg's Nobel Prize–winning 1967 achievement, "A Model of Leptons," received a contemptible five citations between 1967 and 1971 (1967: 0; 1968: 0; 1969: 0; 1970: 1; 1971: 4). It was due to the absence of a means for converting Weinberg's theory into an actual research program. But in 1971 Gerard 't Hooft published a proof showing that a wide range of gauge theories were renormalizable, giving particle physicists the tool they needed to answer various questions that had stood "well-posed" since the 1950s ('t Hooft 2016). Suddenly Weinberg's paper was very popular—it was cited 64 times in 1972 alone; in 1973, 162 times (Galison 1987, 157). By allowing physicists to "carry calculations to any desired degree of accuracy," 't Hooft's proof opened the door to the development of a "comprehensive quantum field theory."[3] What differentiated 't Hooft's proof from work that had come before it was that it offered particle physicists a "way to go on"—a means by which the many, many questions that had long interested them could be satisfactorily resolved.[4]

Thus, the explanatory inadequacy of Kuhn's narrative lies partly in the fact that solving an important problem cannot by itself tell us how to go on, and partly in the fact that the availability of "mop-up" questions offers no guarantee that practitioners will be able to answer them. This is not to deny that the exemplary works on his list played the causal role he attributes to them, nor that they solved important problems, nor that they generated lots of problems for practitioners to work on. My argument is restricted to the conclusion that the promise of future success does not inhere in the fact that an important problem was solved.

In the previous chapter we saw a number of ways in which Kuhn reconsidered many of his important life decisions following *Structure*. These reflections gave rise to several themes

---

[3] Weinberg (1980), 518. Quoted in Gallison (1987), 213.
[4] Many thanks to Cyrus Taylor for patient and detailed discussion of this episode.

that would command Kuhn's attention for the rest of his life. One of them, first appearing in "Postscript," was what he there referred to as "acquired similarity relations," and the central role played by those acquisitions in an individual's ability to attain the concrete successes promised them by paradigm-governed science. Building on his reconceptualization of the status $f = ma$ as an exemplar, he made his first and only attempt to state explicitly the epistemic function of solved problems: "Scientists solve puzzles by modeling them on previous puzzle-solutions" (Kuhn 1962/2012, 189). For post-*Structure* Kuhn, a puzzle-solution is an exemplar that serves as a source of ideas for how to solve other problems. Scientists extract ideas from a previous puzzle-solution "by learning from problems to see situations as like each other, as subjects for the application of the same scientific law or law-sketch" (Kuhn, 1962/2012, 189–190).

Here, as well as in his (1974) "Second Thoughts on Paradigms," he illustrated the process of exemplification by describing a problem-solving lineage beginning with Galileo's study of motion on inclined planes and ending with—of all things—Bernoulli's solution to the problem of the speed of efflux! Galileo describes a similarity between a ball's motion on inclined planes, on the one hand, and the motion of a pendulum, on the other. In the former system, a ball rolling down an inclined plane of any slope returns to the same initial height on a second incline of any slope. He would then draw an analogy between this tendency and that of a pendulum to return to the height from which it descended. In Kuhn's words, Galileo "*learned to see* that experimental situation as like the pendulum with a point-mass for a bob" (Kuhn 1977, 305; my emphasis). Christian Huyghens, upon being presented with the problem of the center of oscillation in 1646 and having "found nothing that made clear even a first approach,"[5] subsequently thought to treat a physical pendulum as composed of many Galilean point pendula,

---

[5] Huyghens, "On the Center of Oscillation," https://www.princeton.edu/~hos/mike/texts/huygens/centosc/huyosc.htm#N_1 (last accessed

whose "collective center of gravity, like that of Galileo's pendulum, would rise only to the height from which the center of gravity of the extended pendulum had begun to fall" (Kuhn 1977, 306). Bernoulli then "*discovered how to make* the flow of water from an orifice in a storage tank *resemble* Huyghens's pendulum" by representing each water particle as a point-pendulum and assuming that the velocity of water leaving a vessel must be sufficient to return the water particles to their original height before descent (Kuhn 1977, 306; my emphasis). If the ascent of the particles' center of gravity must equal the descent of the water as it leaves the tank, then the speed of efflux can be deduced from the speed at which water descends in the tank. In typical Kuhnian fashion, Kuhn's attention is immediately drawn to the causal role of subconscious or perceptual factors involved in treating a previous puzzle-solution as some kind of model for solving other puzzles: "learned to see," "discovered how to make [flow] resemble"—each of these is an instance of "the consequential knowledge of nature acquired while *learning the similarity relationship*" (Kuhn 1962/2012, 190). This way of framing things dovetailed nicely with *Structure*'s depiction of paradigm change as a "shift in vision" over which the practitioner has no rational control, and his more general perception-oriented epistemology of science.

The problem for Kuhn here is that we don't actually have evidence that these episodes involved "learning" or "discovery" of any kind. That language is Kuhn's interpretive overlay, inspired by his belief that the lineage he describes reflects a process that is of a piece with that by which students become adept at solving problems in textbooks: "The student discovers a way to see his problem as like a problem he has already encountered. Once that likeness or analogy has been seen, only manipulative difficulties remain" (Kuhn 1974, 470). He explicitly calls these two examples instances of "the same pattern." Are they, though? The student example involves developmental stages of a single individual, where the appeal to a process of "learning to see" is somewhat plausible. The Galileo example, however, extends over about a century and a half, across several generations of practitioners. It's far from clear what is going on as this lineage unfolds. It *could* be practitioners "learning to see" in the fashion of Kuhn's hypothetical student through the aid of "acquired similarity relations." Or it could be something much more deliberate, much more cognitive, and much less explicable in terms of perceptual experience.

A side-by-side comparison of Kuhn's version of the story with that of Hughyens suffices to illustrate the difficulty. Kuhn says that "Huyghens then solved the problem of the center of oscillation of a physical pendulum *by imagining that the extended body of the latter was composed of Galilean point-pendula*" (Kuhn 1962/2012, 189). What Huyghens actually says is that he "found . . . a way which permits the investigation of this center in lines, surfaces, and solid bodies by a sure method."[6] The difference strikes me as significant. In Kuhn's version, Huyghens appears to visualize the arc of a pendulum anew. In Huyghens's version, he "found . . . a way which permits investigation"—namely, by representing [*referant*] or framing each position on the arc of a pendulum as a distinct pendulum in and of itself, the motion of which is governed by the Principle of *vis viva*: "actual descent equals potential ascent."[7] Where Kuhn saw the all-pervasive influence of perceptual experience, Huyghens seems to describe something more like a *strategy*.

Am I being deliberately uncharitable to Kuhn to artificially inflate the difference between his depiction of Huyghens's experience of inquiry and Huyghens's own account? Yes, of course I am. But it's for a good cause. The point is really just that we simply don't know whether Huyghens was literally "imagining" a particular state of a physical pendulum, or whether Galileo literally "learned to see" a pendular arc as the composite of two inclined planes, or whether Bernoulli believed that efflux "resembled" Huyghens's

---

[6] Huyghens, "On the Center . . ."
[7] Kuhn (1970), 191. This is the principle stated as Hypothesis 1 in Huyghens, "On the Center . . ."

point-pendulum construction. They very well might have just chosen to frame their problem in a certain way to see if they could make any progress. Did it go any further than that? Specifically, did it, as Kuhn claims, make use of "acquired similarity relations" in the manner of the student and his textbook problems? We don't know. What we *do* know is that each man solved his problem by devising a means by which to *represent* it as a modified version of a previous puzzle-solution. Kuhn is correct when he says that these puzzles were solved by "modeling them on previous puzzle-solutions." But he goes too far in crediting their success at modeling to "acquired similarity relations." The appeal to *acquired* similarity relations is a ploy, used to sneak in the desired causal influence of perceptual experience without a warrant.

I am *not* saying that the perception of similarity relations is not something that's acquired; I've had the "textbook problem" experience that Kuhn describes. I also think that there is a lot going for the idea that scientific expertise is very much about acquiring similarity relations that matter for a discipline. But I've also had other kinds of experiences with textbook problems. Painful experiences. Experiences where no similarity relations emerge, even after having done quite a few problems. Yet, on I pressed. In these instances, what I end up doing is *imposing similarity relations by fiat*. I treat *this* part of *this* equation as if it were the same kind of thing as *that* part of *that* equation and see what happens. Sometimes it works. I guess what I'm saying is that there is a more fundamental process afoot here, of which Kuhn's "acquired similarity relations" phenomenon is a special case. That more fundamental process is where— *for reasons left unspecified*—an unsolved problem comes to be treated—*in a manner left unspecified*—as a member of the class to which a solved problem belongs. Each node in Kuhn's "Galileo" problem-solving lineage involves a certain kind of transformation in the epistemic contribution of a particular solved problem, in which the solved problem goes from merely being a solved problem to functioning as a source of ideas for how to approach other problems. This transition begins when a solved problem comes to be seen not just as a solved problem, but as a *sample* of problem solving. Treating the solved problem as a sample of problem solving expands the scope of inferences that we can make on the basis of that solved problem. As a sample of a problem solving, the solved problem tells us not only what that problem's solution is but also— and what is of far greater value—what a problem's solution *can* look like.

With this in mind, let me offer an alternative perspective on the value of a solved problem in the causal process described by Kuhn. In line with the preceding discussion (as well as the previous chapter), let us first observe that the practitioner's choice of whether to work on a given problem will be strongly influenced by whether she believes herself capable of solving it. In this light, we can think of a solved problem—any solved problem—as a member of a certain class of problems we know how to solve. It might be the *only* member; perhaps we know of no other problems like it. Or its relation to other problems may be obvious and intuitive. In either case, that solved problem represents a certain *kind* of problem in the sense that other problems we encounter can be classified with it based on their possession of relevant characteristics. Newton shows that the form of the force law required to generate an ellipse is an inverse-square relation. We know how to solve that problem. But he also shows that an orbit in the form of *any* conic section can be generated by an inverse-square force law. In this example it is easy to see how the problem of elliptical orbits can be classified as one member of a kind of problem involving conical orbits more generally. The problem of conical orbits is a kind of problem we know how to solve; elliptical orbits are members of that kind. Huyghens shows how to find the center of oscillation for a pendulum. But he also shows how the pendulum problem forms part of a very general class of center-of-oscillation problems. We know, in general, how to solve this kind of problem; the pendulum's center of oscillation is a member of that kind.

The promise that some problem can be solved can be derived from its membership in the same class as a solved problem. When some problem can be classified as belonging to the same kind as a problem that has already been solved, practitioners possess a sense—sometimes precise, sometimes vague—of what the solution to their problem will look like. Once their problem has been so classified, it becomes a *solvable problem*. The problem of conical orbits is a kind of problem we know how to solve. When we encounter, say, a hyperbolic orbit, we know how to classify it such that it becomes a solvable problem. Before 1971, one could not perform calculations for interactions between hadronic particles; the integrals were divergent. 't Hooft showed how to combine the relevant Feynman diagrams such that it becomes possible to perform the necessary calculations ('t Hooft 2016). He did not solve all the problems, but he did make them *solvable*. When a problem achieves the status of *solvable*, the practitioner can proceed with confidence and with the understanding that the only obstacle separating her from concrete success is the amount of time and effort typically demanded by a problem of the relevant kind. It will not require the intervention of fate or of genius. A solvable problem promises concrete success to any member of the community with the gumption to tackle it.

Understanding how these classes of solvable problems emerge, and how individual problems become part of those classes, lies at the heart of many of the questions surrounding the nature of fruitfulness and its cognitive role in the growth of scientific knowledge. When a problem is solved, practitioners learn something about the physical or mathematical world. But they also learn that problems "like" this one are *solvable*. In the fully fleshed-out Kuhnian version of this perspective, we would say that the ability of a solved problem to function as a source of solvable problems is foundational to the legitimating role that solved problems play in the development of rational inquiry. When the solution to a particular problem can be treated as a general framework for problem-solving, practitioners have a strong incentive to pursue problems to which that general framework can be applied. Their need for solvable problems will direct them toward its use; as a source of solvable problems, it promises "the concrete successes for which scientists are ordinarily rewarded." The attraction of the exemplary achievements to which Kuhn directs our attention lies not in the fact that they solved an important problem, nor in the fact that they raised follow-up questions, but in their ability to illustrate just how successful inquiry can be when it is conducted within a particular framework. Such sufficiently impressive achievements succeed precisely where mere logical consistency fails: among the many logically permissible ways of proceeding from a solved problem, they provide convincing evidence of the problem-solving effectiveness of a particular set of constraints on inquiry. Practitioners welcome these constraints. Without them, the choice of which path to pursue is effectively random.

The point at which a solved problem comes to be treated as a *sample* of problem solving is the point at which it becomes an *exemplar*. The solved problem need not actually lead to further discoveries or solutions to exemplify problem solving. To function as a sample of problem solving, aspects of the solution need only to be treated as in some way *denoting properties of solved problems*, properties which are possessed by the solution itself in some form or other. It is enough that practitioners see the solved problem merely as a *potential* source of guidance for how to approach other problems, and this they can do simply by treating the solved problem as an example of how one might possibly solve problems. They do not need to acquire similarity relations, as in Kuhn's picture. They can impose them.

Once a solved problem is seen as a possible guide to problem solving, practitioners begin looking for ways of making connections to other problems. They routinely find themselves in the position aptly described in Huyghens's lament at having "found nothing that made clear even a first approach" to the center of oscillation. With nothing to recommend one approach over another, the costs of investing in such a problem will hardly seem justified; there is no promise of concrete success. A solved problem—often in an entirely distinct field of inquiry—will entice practitioners to see whether it can help to provide a first glimpse of how their own unsolved problem might be approached. The solved problem has now begun to helpfully constrain inquiry by highlighting one promising path among the many logically possible ways of proceeding: Aspects of motion on inclined planes suggested to Galileo how one might analyze the motion of a pendulum. His solution to pendular motion inspired Huyghens to approach centers of oscillation from the perspective of pendular motion. Bernoulli successfully sought to apply aspects of Huyghens's solution to the problem of efflux. When practitioners describe something as having been fruitful, this is often the sort of cascade they have in mind: a discovery or innovation that begets more "concrete successes." In these cases, something about the way in which an initial problem is solved gives practitioners a glimpse of how to approach another problem. That glimpse may be a reflexive perception of a similarity between the solved problem and the unsolved problem, as perhaps in Kuhn's description of Galileo as having "*learned to see*" a rolling ball "as like" a pendulum. Or it may be the result of a deliberate attempt to creatively link a solved and unsolved problem together. Both involve the act of *framing* a problem such that it becomes a member of a solvable class. For example, classical mechanics treats an explosion as a time-reversed perfectly inelastic collision. I would submit that the classification of an explosion as a species of perfectly inelastic collision did not arise as the result of someone "learning to see" an explosion as a collision running backward through time. Rather, we represent an explosion as a species of perfectly inelastic collision because the analysis of perfectly inelastic collisions is a relatively simple exercise that allows us to ignore things like the direction of time. We wanted to represent an explosion as a species of collision, and so we did. Even if we only look in the area illuminated by the lamppost, the inventory of things in that area needn't remain fixed. We are very good at dragging things into the light.

3.4. Problems Structured

Up to this point, we've mostly been concerned with developing a more fully articulated *Kuhnian* picture of where the promise of concrete success comes from. Much of the time, this has involved standing on Kuhn's shoulders and pausing to focus on details which would have complicated his grand vision but which are essential to an epistemologically coherent picture of scientific decision-making. Other times it has involved stepping on Kuhn's face in a pathetic bid for self-aggrandizement. This is going to be one of those other times.

Kuhn's explanatory use of great works of science derives its power from the double life that each of them was fortunate enough to have led: they all involve both (1) the solution to an important scientific problem as well as (2) the illustration of an approach to structuring scientific investigations. Book I of Newton's *Principia* was understood to solve, *inter alia*, the pressing seventeenth-century problem of what form a force would have to take to produce an elliptical orbit. Book I also displays the limitless power of Euclidean geometry for structuring problems in mechanics. The problem of elliptical orbits is not just a solved problem. It is an example of how to structure problems in mechanics such that the forces involved can be calculated. It's literally a tour de force. (No, I'm not sorry I said that.) I will assume that it is clear enough how solving an important problem can cause a stir. Often enough, though, we find that researchers are similarly inspired by attempts that are *not* widely accepted as successful. That is, even if practitioners do not regard the particular research problem as having been solved, they nevertheless draw inspiration from the *attempt itself*. In the simplest case, a practitioner might approach the same research problem using a slightly refined version of the original attempt, such as when Newton adds a mass term to Kepler's Harmonic Law to increase its accuracy (see, e.g., Cohen 1983, chapter 5). Newton saw as essentially correct *in spirit* Kepler's solution to the relationship between a planet's mean distance from the sun and its period of revolution, but he also saw reasons for thinking that a better solution might be obtained by considering the effect of a planet's mass.

We know that this sort of development is routine. Kuhn presumably did as well. Although he cites Truesdell's (1967) study of the reception of Newton's work on fluid dynamics as support for his claim that "mechanics progressed . . . by modeling one problem-solution on another" (Kuhn 1962/2012, 189 fn. 11), much of the focus of Truesdell's paper is on how practitioners largely rejected Newton's solutions to the problems in fluid dynamics. That is, they did not regard them as solutions at all. And yet, Book II of the *Principia* was incontestably revolutionary, for it gave practitioners a clear vision for how to "frame" or "structure" the problems in fluid dynamics such that they *could* eventually arrive at solutions that garnered the assent of the community. A failed attempt to completely solve a specific research problem to the satisfaction of the research community as a whole nevertheless pointed the way to an attempt around which consensus eventually did form. This can happen within the span of a single career, such as when a practitioner's earlier unsatisfactory efforts lay the foundation for the eventual solution to the problem. It can happen within a single generation of a given research community. Or, as with Newton's emendation of Kepler, a problem-solving attempt might garner consensus during one generation but come to be seen by later generations (often due to improvements in data or technology) as roughly right, though not quite satisfactory. And, of course, there will be instances in which one part of the research community takes a problem to be solved while another part of the community regards that same problem as still awaiting resolution.

The fluid and fickle behavior of a problem's status as "solved" suggests an important lesson, and brings into sharp relief the fundamental weakness in Kuhn's narrative of how exemplars emerge: occasions on which a solved problem inspires attempts to approach other problems in a related way might be best understood as a special case of the more general phenomenon wherein an *attempt* to solve a problem generates related efforts. That is, the historical phenomenon of interest seems to be one in which a problem-solving attempt per se inspires a host of related problem-solving attempts; sometimes the original attempt is accepted as a successful solution; sometimes not. Whether or not the research community sees the research problem as solved, practitioners are able to use the attempt to guide future research efforts. They are able to do this by treating the attempt as a sample of problem solving— specifically, as an exemplification of an approach to "framing" or "structuring" a problem.

The notion that successfully solving an outstanding problem is not necessary for a problem-solving effort to inspire related attempts is perhaps best illustrated by instances in which a new problem-solving method is explicitly introduced as an alternative to existing methods. These introductions typically take the form of showing how the new method is able to obtain results that are already known to be correct. Perhaps the method makes results easier to calculate, as with Feynman diagrams (Kaiser 2005). Sometimes the new method provides a simpler or more explanatory proof of a theorem that had first been proven long ago, as with Fermat's analytic method for determining loci for curves that had been conquered by ancient geometers through unknown means (Mahoney 1973, chapter 2). Sometimes the point is just to show that the method *works*, not that it is necessarily better than existing tools. Whatever the new method's alleged virtues are, they are most convincingly illustrated by applying them in contexts where there is no doubt as to the problem's solution. There can be no question that, in such cases, the new problem-solving approach's ability to inspire related efforts does not depend on solving an outstanding research problem. Of course, the hope is often that it will aid in the solution of outstanding problems, and sometimes the best way to illustrate their promise is by showing that it gets the right answer to other problems. In this sort of case, the problem-solving illustrations are treated explicitly as samples of the new method's ability to solve *certain kinds of problems*, with the intention that readers appreciate the illustrations as such (Why else report a result that readers are expected to already know?). The problems are themselves chosen for their exemplary nature—that is, for their own ability to serve as samples of certain kinds of problems. When we are shown that the new method gets the correct answer to a problem that exemplifies a *kind* of problem, we are intended to infer that the new method is generally effective in these familiar problem-solving contexts. This is the approach taken by Newton in his 1666 demonstration of the method of fluxions, for instance (Westfall 1980, 134–135). It is the approach taken by Fermat to illustrate the power of his loci technique. Readers were not uniformly adept at following his lead.

But the intended inferences are typically much more far-reaching than this. For, in showing the reader that the new approach is capable of solving a certain kind (or several different kinds) of problem, an argument has in effect been made that, together, the problems that can be solved with the new approach *themselves* constitute samples of a new kind of problem—viz., the kind of problem that can be solved with the new method. The problem-solving illustrations show that, by *structuring* problems in a certain way, we will be reliably led to solutions. And herein lies the promise of the new method: the problem-solving illustrations have revealed a hitherto unknown class of solvable problems.

Let's take stock. In what I imagine would have been the fully elaborated version of Kuhn's narrative of exemplar emergence, the promise of concrete successes is established by the emergence of a class of solvable problems. Kuhn assumed that such classes emerge on the back of solved problems. But this is too restrictive. Although

Kuhn's solved problems are exemplars, what they exemplify is a *kind of problem*. A kind of problem is characterized by a certain structure. These kinds will emerge under any conditions in which a problem-solving attempt is treated as a sample of how to make problems solvable. This can occur (1) when the problem-solving attempt successfully solves an outstanding problem (Kuhn's version); (2) when the problem-solving attempt is unsuccessful in its effort to solve an outstanding problem; or (3) when the problem-solving attempt approaches in a new way a problem that has already been solved. Because successfully solving an outstanding research problem is only one of a range of contexts in which the promise of a new approach is established, it cannot be considered essential for establishing promise. Rather, the most plausible explanation for why the successful solution to an outstanding problem is often accompanied by the perception of promise is that solving an outstanding problem, like sex in advertising "doesn't sell, it just gathers a crowd" (Brush 2015, 346).

The same could be said for problem-solving attempts that were not generally accepted: a plausible, novel, but not quite satisfactory attempt often gathers a crowd of curious, interested, or desperate researchers in search of a way of pushing inquiry forward. This would explain why failed attempts to solve outstanding problems can have the same effect as novel approaches to solved problems. It would also explain why, in the case where a new approach is *not* introduced to solve an outstanding problem, there is frequently a significant lag time between when the publication of the approach and its eventual application to new research problems. If attempting to solve an outstanding problem is what "gathers a crowd," and no attempt to solve an outstanding problem has been made, the publication of the new approach will lack a powerful mechanism for attracting the attention of researchers. This latter phenomenon is easily discernible in the historical relationship between mathematics and physical science, where mathematical techniques—such as differential geometry and Galois theory—often go unnoticed for decades or more before a research problem arises which resists efforts to apply the mathematical tools that are in circulation at the time. At some point, often through chance encounters, physicists realize that a particularly recalcitrant outstanding physical problem can be represented as a member of an established family of problems in mathematics, thus making the problem solvable *per* our earlier definition.

This leads us back to Kuhn. It is worth pausing to reflect on the likely prospect that what attracts researchers away from competing modes of scientific activity is probably *not* the same thing that keeps them enduringly attached to a research framework. Researchers will be induced out of hope, skepticism, or just plain old curiosity to examine a new approach to framing a research problem of widely acknowledged import. But whether they use the new approach to in some way help guide their own research will depend on *whether they are able to envision some way in which aspects of the new approach could be extended to other research problems.* The mere fact that the new approach solves an outstanding problem cannot tell them this; something else must be involved. That something else, I will argue, is a "structure" that characterizes the kind of problem exemplified by the problem-solving attempt.

3.5. Recap, Conclusion, and Coming Attractions

We began this perilous journey by generously presuming that there is something correct about the claim made in *Structure* that paradigm-governed science emerges in response to someone's having solved an important research problem. Looking at some of Kuhn's examples, both in *Structure* and beyond, I tried to supply some of the details for how his version of this process might work. That led to the hypothesis that, in Kuhn's examples, a solved problem comes to be treated as *a sample of how problems "like this" can be solved*. When treated as such, the solved problem denotes a certain class of solvable problems. I think that, with a sufficient but not unreasonable amount of charity, this view can be attributed to Thomas Kuhn.

I then argued that this view suffers from two critical defects. The first defect is that it holds that solved problems alone possess the power to denote a class of solvable problems. As such, the promise of future concrete successes should inhere only in solved problems. This runs counter to the historical record in various ways. For one thing, the promise of concrete success has frequently been perceived in *failed* attempts to solve a problem. Also, later generations often reject the solutions of their predecessors, while benefiting from the actualization of the promise those apparent solutions did, in fact, portend. Thus, the *solution* to a problem cannot be what practitioners take as denotative of a class of solvable problems.

The second critical defect is that, on Kuhn's view, the criteria for class membership are implicitly defined by the way a class is deployed in the practice of science, experience with which results in the acquisition of the implicitly defined criteria; these are Kuhn's "acquired similarity relations." A problem becomes a member of a solvable class through its possession of these implicitly defined criteria. Once again, the problem for Kuhn's view here is that this is too restrictive as an account of how approaches to problems in science are modeled on previous approaches. Even his own examples do not speak firmly in favor of his "acquired similarity relations" perspective. Kuhn wanted the way we classify experience—"world-making"—to be rooted in base perceptual capacities. And he wanted the scientist's taxonomic practice to be rooted in the perceptual capacities she acquires as part of the practice of science. Some of her taxonomic practices surely fall into this category. Intensive study and observation can affect perceptual experience and awareness in precisely the manner Kuhn envisions. A radiologist's perceptual experience of an X-ray is just different from mine. And some of her taxonomic practices are just as surely the result of similarity relations acquired *before* her life as a practicing scientist; it is not in virtue of her rigorous scientific training that she assigns objects to the class *elephant*. The acquisition of these similarity relations may often go hand in hand with the acquisition of a language, as Kuhn increasingly came to believe (beginning with Kuhn 1974—his commitment steadily increases thereafter). But the reality of the classificatory phenomenon Kuhn describes must not be mistaken for the generality with which it applies. His notion that puzzle-solving lineages are made possible by the fact that problems are bound together through acquired similarity relations that inform perceptual experience needlessly limits the creative freedom of scientists to try out by fiat different ways of binding problems together in a single class, pursuing alternate strategies for classifying problems when others perform unsatisfactorily. We saw instances of scientists exercising this freedom. It is hard to imagine why they wouldn't. So, while we should not deny that scientists can "come to see" (in Kuhn's thoroughly perceptual sense) a pendulum as a system of inclined planes, I hope to show that the normal sequence of events is very much the reverse of how he conceived of them. Instead of similar approaches being a consequence of having come to see problems as alike, scientists come to see one problem as like another as a *consequence* of the decision to treat them similarly.

The remedy for both defects involves the transition to a perspective according to which membership in a class of solvable problems is a matter not of having solved a problem, but rather of an approach to the set-up or structuring of problems. Sometimes these problem set-ups or structures will emerge intuitively out of experience, as is suggested by Kuhn's example of learning how to do textbook problems. But they won't always do that. This does not mean that the set-up or structuring of problems is thereby rendered impossible. It is not only possible; it is routine. Nor does it mean that any attempt at setting up or structuring problems will produce worse outcomes than it otherwise might have, had it descended from acquired similarity relations.

This chapter has been devoted to the search for a perceptible source of promise, a search necessitated by Kuhn's stubborn assertion that faith was the only possible foundation upon which to base decisions about future promise. His hopes for knee-capping the rationality of such decisions rested on insisting that they are made "in defiance of the evidence provided by problem solving." This is somewhat ironic; for all Kuhn's opposition to the dominant epistemologies of the day, it would be hard to find a more full-throated endorsement of them. With the benefit of sixty years of hindsight, I feel confident in the view that the philosophical reception of *Structure* was not well handled. And I feel equally confident in saying that most of the bungling was due to philosophers' pronounced lack of concern for how science actually works, combined with their presumption that it must take whatever form we've imagined it to take in our most pleasant epistemological dreams. The publication of *Structure* was a huge missed opportunity. For all the damage their reception did, however, the one thing they got right was that Kuhn did not understand what his views implied for the epistemology of science. He thought that philosophers were deliberately misinterpreting him. Fair enough—some of us do that sometimes. Also, we are pointlessly critical (except for me; *my* abuse of Kuhn is totally warranted). But it does seem that everyone except Kuhn understood *Structure* to be a clear rejection of the idea that scientific decision-making is rational in an epistemically meaningful sense. So you tell me who's crazy.

In defiance of the evidence provided by Kuhn, I think we have found the basis for perceptions of the promise of future concrete successes. Kuhn was right about one thing (well, a lot of things, actually): promise does not inhere in the evidence provided by problem solving. It inheres in the evidence provided by problem *structuring*. Unlike knowing whether a scientific problem is solved, individual practitioners can know more or less instantaneously whether a problem is set up in such a way that they can work on it. This difference alone tells us something of profound importance. It explains how whole scientific communities can proceed both *rapidly* and *rationally* toward the adoption of a new paradigm, in advance—even in defiance—of the evidence provided by problem solving; not just when one paradigm "has succeeded only with a few," but even when it has succeeded with none. It also explains the legitimating effects of exemplary works of science to which Kuhn famously drew attention. When we know that problems of a certain kind—characterized by a certain set-up or structure—are solvable, we have license to work on problems of that kind.

The remaining chapters of our study are devoted to filling out this alternative epistemological perspective in full detail. What *are* these "set-ups," "structures," or "frames" that are embodied by problem-solving attempts and that are supposed to be the real locus of the promise that rationally motivates practitioners' decisions regarding what to work on? How do those structures give rise to entire classes or kinds of problems? How does the emergence of a new kind of problem precipitate the familiar cascade of concrete scientific achievements traditionally associated with the notion of fruitfulness? How do these problem structures serve as the foundation for the alternate species of epistemic warrant of which we are now in search? And what is it that facilitates the superior performance of some problem structures over others as a foundation for warrant, such that scientists might rationally choose between them?

4

The Emergence of Exemplars

In late 1668, Isaac Barrow, then the inaugural Lucasian Professor of Mathematics at Cambridge, received a copy of Nicolaus Mercator's recently published *Logarithmotechnica*. Most significantly, Mercator's book described a simplified means for determining the area under the hyperbola. His description begins in the third and final section by indicating that the equation for a hyperbola $1/(1 + x)$ can be written as the infinite series $1 - x + x^2 - x^3 + x^4 \ldots$ (see Figs. 4.1 and 4.2). He then used this result to suggest that the area under a given section of the hyperbola can be determined by adding together the magnitude of every line connecting the hyperbola to the abscissa across the specified section (Mercator 1668, 32). Were one to be so bold, she would find that the area is given by the series $x - x^2/2 + x^3/3$ &c., which Mercator did not write down but which he demonstrated by showing the series calculations for a couple of segments. Barrow immediately wrote to Newton to apprise him of Mercator's discovery.

What was it that Barrow saw in *Logarithmotechnica* that made him think of Newton? Why, Newton had shown him the same result after finding it himself a few years earlier—probably in the first months of 1665 but possibly as early as winter 1664—while an isolated B.A. student at Cambridge.[1] But he had found considerably more than that (see Fig. 4.3). Whereas Mercator had written down the expansion of a single binomial $(1 + x)^{-1}$, Newton had quickly parlayed his own result into the general binomial theorem a few months later, when he realized that the coefficients for any

---

[1] See Newton and Whiteside (1967, 113).

Fruitfulness. Chris Haufe, Oxford University Press. © Oxford University Press 2024.
DOI: 10.1093/oso/9780197666395.003.0004

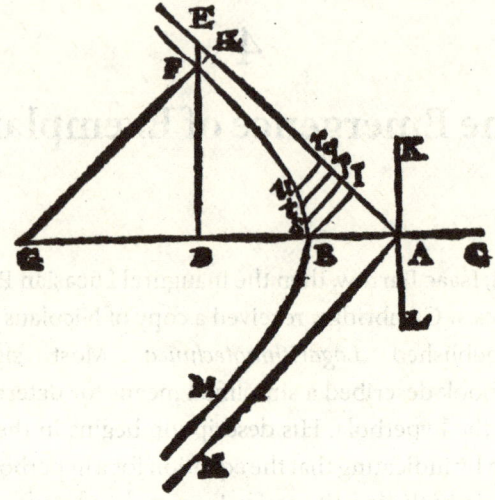

Figure 4.1 Mercator's quadrature of the hyperbola (1668, 28).

Demonftratio.   Sit $AB = 1 = AC$
$BC = AB + AC = 2$
$BD = a$
$AD = AB + BD = 1 + a = DE$
$CD = BC + BD = 2 + a$
$CD * BD = 2 + a$ in $a = 2 a + aa = Q : DF$
$DF = \sqrt{2a + aa} = DG$
$AG = AD + DG = 1 + a + \sqrt{2a + aa}$
$\sqrt{2} . 1 :: AG . AH$
$$AH = \frac{1 + a + \sqrt{2a + aa}}{\sqrt{2}}$$
$EF = DE - DF = 1 + a - \sqrt{2a + aa}$
$\sqrt{2} . 1 :: EF . FH$
$$FH = \frac{1 + a - \sqrt{2a + aa}}{\sqrt{2}}$$
Ducatur AH in FH
ponendo $1 + a = c$, & $\sqrt{2a + aa} = d$
erit $1 + 2a + aa = cc$ } fubtrahe
& $2a + aa = dd$ }
$$\frac{1}{} = cc - dd$$
$$\frac{cc - dd}{\sqrt{2} * \sqrt{2}} = \frac{1}{2} = AH * FH$$
$\sqrt{2} . 1 :: AB . AI$
$$AI = \frac{1}{\sqrt{2}}$$
$$AI * BI = \frac{1}{\sqrt{2}} * \frac{1}{\sqrt{2}} = \frac{1}{2}$$ β
Ergo per α & β, $AH * FH = AI * BI$
& $AH . AI :: BI FH$ q.e.d.

Figure 4.2 Mercator's derivation of the area (1668, 28–29).

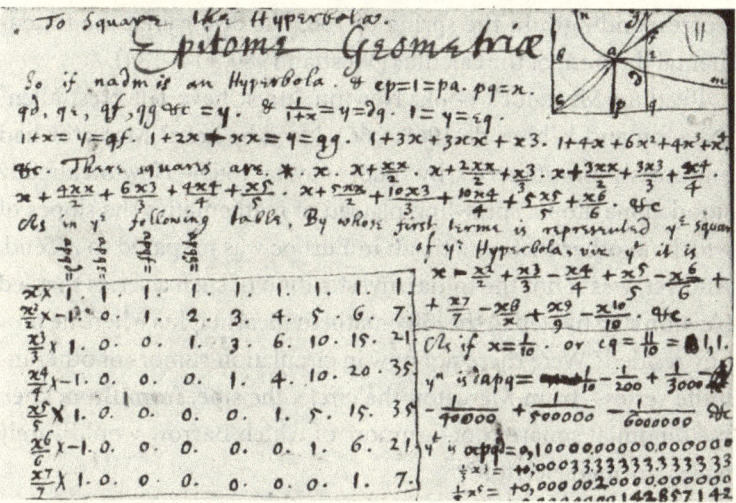

Figure 4.3 Newton's derivation of the Mercator series, center right. MS. Add. 4000 f. 20r.

Figure 4.4 Newton's derivation of the general binomial expansion. MS Add. 3958.3, f. 71.

binomial of the form $(b + x)^{m/n}$ could be generated by the algorithm $[1 \times m \times (m - n) \times (m - 2n) \times (m - 3n) \times (m - 4n)]/(1 \times n \times 2n \times 3n \times 4n \times 5n)$.[2] (See Fig. 4.4.) Just as Mercator had done, Newton had connected the expansion of the binomial to the problem of calculating the area under a curve; but since Newton's approach was general, so too was the scope of his ability to calculate areas.

---

[2] Whiteside (1961). See Newton and Whiteside (1967, 126–133), which contains the notes from Newton's manuscript *"De Solutione Problematic per Motum"* (MS. Add 3958.3 f71).

Correspondingly, in the spring of 1665, he discovered the fundamental theorem of the calculus (Westfall 1980, 114–134).

Reading Mercator's book, Newton "must have felt crestfallen" (Newton and Whiteside 1968, 166). Need he have? Mercator had one measly infinite series; by 1668—nay, by 1665—Newton already stood alone atop a sprawling plateau of mathematics the slopes of which no other mathematician in Europe was prepared to ascend. And yet, was it not the initial investigation of such a series that led Newton step by step to the lofty mathematical heights where he now idly resided? Were there not now in circulation rumors of other infinite series—from Mercator, the circle, the sine; from Brouckner, the binomial square root—rumors of which Barrow would surely have informed Newton?[3]

> Mercator's exposition of the procedure was admittedly cumbrous and inadequate, but if he could employ Newton's reduction to infinite series *dividendo* in a particular case, how long would it be before he stumbled on the extraction of roots in such series and indeed upon his cherished binomial expansion? . . . Gradually it became clear that Newton could choose either to circulate an account of his own discoveries in infinite series or to relinquish the credit for those discoveries to whoever published the inevitable complement to Mercator's book. (Newton and Whiteside 1967, 166–167)

Barrow's note, and Newton's essentially instantaneous response in the form of *De Analysi*—a refined exposition of his method of series[4]—suggest that they agreed that able mathematicians could not fail to see Mercator's particular series as more than just the integral for the hyperbola $1/(1 + x)$ defined over the interval 0–1,

---

[3] Newton and Whiteside (1968, 167).
[4] Whiteside writes that "[p]robably it was composed over a period of a few days" (Newton and Whiteside 1968, 206 fn 1).

and as more than just the series for log (1 + x) (which Newton and Mercator had also seen). This particular series would, they knew, be treated as but a sample of what the future of mathematics promised.

The object of our attention in this chapter is the kind of experience of which Barrow, Newton, Mercator, and various other mid-seventeenth century mathematicians partook, the kind of experience that gave rise to Mercator's discovery, that prompted Barrow to write to Newton, that gave Newton great concern, and that established him as the foremost mathematician in Europe. It is the experience of treating a problem-solving attempt as a sample of problem solving. This is the kind of experience associated with exemplification.

4.1. "All the Lines"

In truth, there was more than just the rediscovery of his binomial expansion that would have given Newton a jolt; Mercator's approach to the problem reflected disturbingly similar conceptual foundations. It appears that he and Newton had both been taking rips off the same mid-seventeenth-century mathematical bong—an arithmetical version of the "quadrature" method for determining the area under a curve by dividing it up into sections whose areas could be calculated. Known side effects included a tendency to believe that one had the ability to add up arbitrarily small magnitudes (see Figs. 4.5 and 4.6).

A confluence of mathematical events in the first half of the seventeenth century had created an intellectual environment in which it became possible to view the ancient Greek geometrical approach to measuring areas through the lens of arithmetic. The first of these was the "method of indivisibles" for which Cavalieri's 1635 *Geometria indivisibilibus continuorum nova quadam ratione promota* ("Geometry, advanced in a new way by the indivisibles of the continua") is traditionally considered the point of origin.

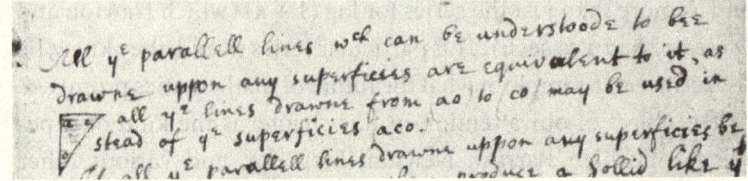

Figure 4.5 Newton's conception of a surface as the sum of its parallel lines, 1664. MS Add. 4000 82r.

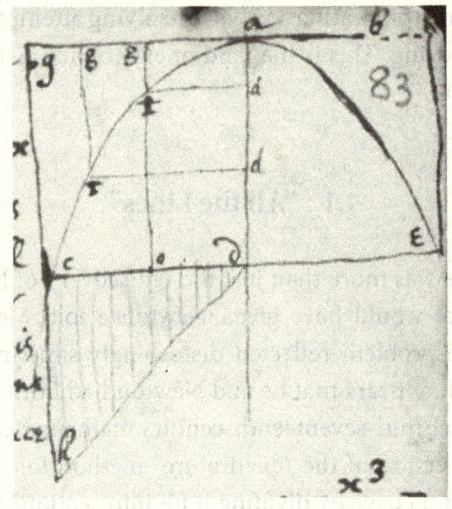

Figure 4.6 Newton's diagram for the quadrature of the parabola, 1664. MS Add. 4000 83r.

The principal contribution of this work to later mathematics was to promote the conceptualization of a plane figure as being characterized by the collection of lines generated by the uniform motion of a straight line from one end of the plane to the other (Andersen 1985, 300–301). The "motion" of points, lines, and planes was an increasingly common conceptual device in seventeenth-century mathematics; to its credit belong several groundbreaking mathematical insights (Mahoney 1998,

724–726). But its effect on the ancient art of quadrature seems to have been particularly generative. Perhaps by visualizing the sort of change a line's magnitude would have to undergo as it moved indefinitely along the abscissa underneath the hyperbola, or as it approached the ordinate beneath a parabola, its inevitable descent into "infinite littleness" became perceptible as never before. We know that Mercator heard lectures on Cavalieri while in Copenhagen (Hofmann 1939, fn9). It was Cavalieri's student Evangelista Torricelli who, a short while later (1644), treated Cavalieri's collections of lines as defining infinitesimally thin rectangles whose sum constituted the area of the figure (Andersen 1985, 355). Torricelli's work was widely read and appreciated, and proved to be an attractive entry point for mathematicians across Europe to start thinking about the mathematics of sums of infinitesimals (Stedall 2005, 25).

The other key foundational development that arose during this time was Descartes's *Geometrie*, which was the first published work to systematically treat the curves of classical geometry as sets of points that could be described by algebraic equations. It is not possible to overstate the effect that this book had on the subsequent evolution of mathematics. To name just one, it allowed for the arithmetical manipulation of all sorts of properties of curves, freeing mathematicians from certain weighty constraints imposed on them by proportional geometry. In addition to fundamentally transforming the practice of geometry, it served as something like an ambassador for algebra itself, which was still in seventeenth-century Europe somewhat distrusted as a mathematical. Much like Cavalieri's effort to frame planar figures as collections of lines (or solid figures as collections of planes), Descartes's algebraic description of curves would transform the very manner in which those curves were conceived by subsequent generations of mathematicians.

Newton and Mercator were part of the first generation to be reared in this new conceptual milieu, and what one sees in their work and that of other mathematicians of that time is the emergence of a new perspective on geometry that resulted from the overlapping of two mathematical domains that had traditionally been kept apart. The juxtaposition of these different ways of representing magnitudes flicked the same switch in the minds of several mathematicians (Mahoney 1998). Once that switch was on, it appears to have illuminated a number of well-defined pathways for inquiry, of which the Newton-Mercator approach to the hyperbola was but a sample; other mathematicians independently set out on neighboring roads. These pathways were defined by an approach to the study of areas (and volumes) that allowed mathematicians to treat traditional problems in geometry as problems of arithmetic, in that it provided them with a way to set up those traditional problems as algebraic equations rather than proportions. The works of Cavalieri, Torricelli, and Descartes together served as a sample of how one could approach those traditional problems. Their audiences applied strikingly similar interpretations of such an approach to the treatment other problems not explicitly taken up by these authors.

4.2. Dr. Wallis and the Infinity Gauntlet

The principal watershed moment for the arithmetical framing of problems of area and volume was the publication of the *Arithmetica Infinitorum* (1656) by John Wallis, which really took this emerging perspective to the limit (not sorry I said that either). In it, Wallis framed the problem of quadrature in explicitly arithmetical terms and traced out a general arithmetical pattern for depicting infinite series. While his approach was by no means free of the basic problems that had plagued infinitesimals from Day 1, it doesn't really seem to have mattered to the approach's ability to "prepare the ground for some of the astonishing advances of the second half [of the seventeenth century]: the discovery of the general binomial theorem, applications of infinite series, and the integral calculus" (Stedall 2005, 24). Wallis's approach was able to recover enough results already known to be correct that it encouraged the perception that it generally worked, and could be extended and adapted at will to the needs of mathematicians so long as they weren't too fussy about the strict mathematical or metaphysical coherence of the foundations. Newton's debt to Wallis is by now platitudinous. Our purpose in this section will be to highlight certain notable aspects Wallis's approach which make that debt difficult to square with the Kuhnian version of exemplars as solved problems, but easy to square with the idea of exemplars as promising set-ups (i.e., my idea).

The signature contribution of *Arithmetica Infinitorum* was to suggest a method for calculating areas through the use of infinite series in the context of the quadrature of the parabola (see Figs. 4.7 and 4.8).

We want to find the area of the section *ATO*—more precisely, we want to find the ratio of its area to that of the rectangle *ATOD* (see Figs. 4.9 and 4.10).

1. Following Toriccelli, Wallis conceives of *ATO*'s area as equivalent to the sum of all the ordinates in *ATO* parallel to *TO*, and the area of rectangle *ATOD* as equivalent to the sum of all the ordinates parallel to *TO* in that rectangle.
2. Now, because the parabola is given by $y = x^2$, each *TO* is equivalent to $(OD)^2$ (think of the length of *OD* as the x-value, and *TO* as the y-value).
3. Take the first ordinate in *ATO* to have length 0^2, the second to have length 1^2, the third to have 2^2, and so on such that the nth ordinate (*TO*) has a length of n^2.
4. The area of *ATO* can then be represented as the sum of all those lines $0 + 1 + 4 + 9 + \ldots + n^2$, while the area of *ATOD* can be represented as the sum of all *its* lines, each of which has length n^2.
5. Thus, the ratio of the areas of *ATO* to *ATOD* will be $(1/3) + (1/6n)$.

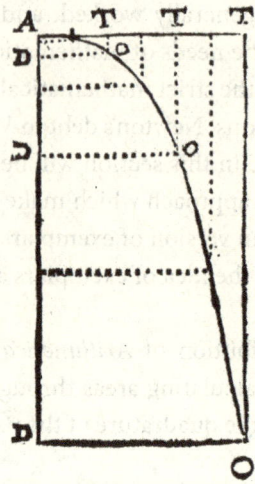

Figure 4.7 Wallis's quadrature of the parabola (1656, 45).

Fiat inveſtigatio per modum inductionis, (ut in prop. 1.)

$$\text{erĸq;} \quad \frac{0+1=1}{1+1=2} = \frac{1}{6} = \frac{1}{3} + \frac{1}{6}. \qquad \frac{0+1+4=5}{4+4+4=12} = \frac{1}{3} + \frac{1}{12}.$$

$$\frac{0+1+4+9=14}{9+9+9+9=36} = \frac{7}{18} = \frac{1}{3} + \frac{1}{18}. \qquad \frac{0+1+4+9+16=30}{16+16+16+16+16=80} = \frac{3}{8} =$$

$$\frac{2}{64} = \frac{1}{3} + \frac{1}{24}. \qquad \frac{0+1+4+9+16+25=55}{25+25+25+25+25+25=150} = \frac{11}{30} = \frac{1}{3} + \frac{1}{30}.$$

$$\frac{0+1+4+9+16+25+36=91}{36+36+36+36+36+36+36=252} = \frac{13}{36} = \frac{1}{3} + \frac{1}{36}.$$

Figure 4.8 Wallis's proof (1656, 15–16). The text contains a handful of such demonstrations for other figures, never going beyond a few sample calculations.

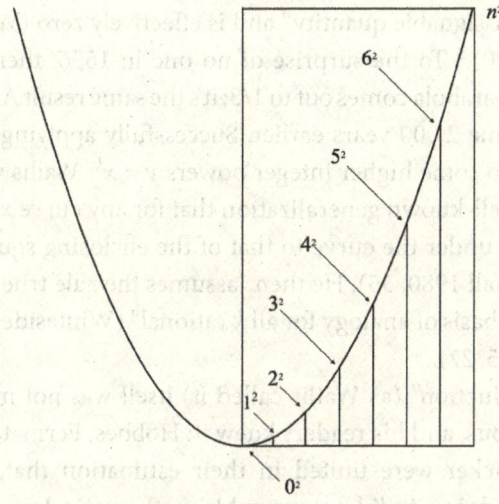

Figure 4.9 Quadrature of the parabola.

$$\frac{0^2 + 1^2}{1^2 + 1^2} \qquad\qquad \frac{1}{3} + \frac{1}{6} = \frac{1}{3} + \frac{1}{6 \cdot 1}$$

$$\frac{0^2 + 1^2 + 2^2}{2^2 + 2^2 + 2^2} \qquad\qquad \frac{5}{12} = \frac{1}{3} + \frac{1}{6 \cdot 2}$$

$$\frac{0^2 + 1^2 + 2^2 + 3^2}{3^2 + 3^2 + 3^2 + 3^2} \qquad\qquad \frac{14}{36} = \frac{1}{3} + \frac{1}{6 \cdot 3}$$

$$\frac{0^2 + 1^2 + 2^2 + 3^2 + \ldots + n^2}{n^2 + n^2 + n^2 + n^2 + \ldots + n^2} \qquad\qquad \frac{1}{3} + \frac{1}{6n}$$

Figure 4.10 Quadrature of the parabola.

Wallis then remarked on the fact that, for the cases he looked at, each ratio can be written as 1/3 plus some additional fraction 1/6n that gets increasingly small as the number of n-intervals goes up. He asserted that, ultimately, the additional fraction becomes smaller than any "assignable quantity" and is effectively zero (Wallis 1656, 16, Prop. 20).[5] To the surprise of no one in 1656, then, the area under the parabola comes out to 1/3; it's the same result Archimedes derived some 2,000 years earlier. Successfully applying this same approach to some higher integer powers $y = x^k$, Wallis went on to infer the well-known generalization that for any curve x^k, the ratio of the area under the curve to that of the enclosing square is $1/(k + 1)$ (Westfall 1980, 35). He then "assumes the rule true on a mere instinctive basis of analogy for all k rational" (Whiteside 1961, 237; Stedall 2005, 27).

The "induction" (as Wallis called it) itself was not mathematically rigorous, and his readers knew it; Hobbes, Fermat, Huygens, and Brouncker were united in their estimation that, whatever Wallis was doing, it did not resemble mathematical proof (Stedall 2005, 29). Moreover, Wallis's appeal to infinity was never free from the familiar conceptual hang-ups associated with infinitesimals. Here is Hobbes, displaying his distinctive knack for resisting conceptual developments that would quickly come to define modernity:

> "The triangle consists as it were" ("as it were" is no phrase of a geometrician) "of an infinite number of straight lines." Does it so? Then by your own doctrine, which is, that "lines have no breadth," the altitude of your triangle consisteth of an infinite number of "no altitudes," that is of an infinite number of nothings, and consequently the area of your triangle has no quantity. If you say that by the parallels you mean infinitely little parallelograms, you are never the better; for if infinitely little, either they are nothing, or if somewhat, yet seeing that no two sides of a triangle are parallel, those parallels cannot be parallelograms. (Hobbes 1656, 46; quoted in Stedall 2005, 30)

---

[5] My exposition follows Stedall (2005).

In sum, Wallis's contemporaries were as unconvinced by his demonstrations—which did not conform to prevailing standards of rigor—as they were unimpressed by his results—which they mostly already knew.

If we focus on his lack of demonstrative rigor or the incoherence of his conceptual foundations, though, we fail to appreciate the enormity of Wallis's contribution. As the seventeenth century progressed, the virtues of arithmetic were gradually becoming more and more apparent. By presenting the fundamental details of an alternative form of mathematical reasoning—a new, arithmetical way of setting up old problems—and by indicating its plausibility through the derivation of results known to be correct, Wallis invited similarly inclined mathematicians to investigate more fully the merits of an arithmetical approach. While mostly unconvinced by his specific demonstrations, readers nevertheless saw *something* in *Arithmetica Infinitorum* that they believed held promise for future investigations. Indeed, this was an experience shared by Wallis himself. Responding to Fermat, Wallis defended *Arithmetica Infinitorum* as exemplifying a method of *investigation*, rather than mathematical demonstration:

> [Fermat] doth wholly mistake the design of that Treatise; which was not so much to shew a Method of Demonstrating things already known [. . .] as to shew a way of *Investigation* or finding out of things yet unknown [. . .] and that therefore I rather deserved thanks, than blame, when I did not only prove to be true what I had found out; but shewed also, how I found it, and how others might (by those Methods) find the like. (Wallis 1685, 305–306. Quoted in Stedall 2005, 29; emphasis is in Wallis)

Wallis's advance lay in the development an *arithmetical approach*—rather than a geometric proof—that could reproduce results known to be correct, thus recommending "how others might (by those Methods) find the like." The essence of his attack on these and related problems consisted of the rigorous derivation of a few results, followed by a generalization which he took to be justified "by induction," or "by analogy."

Remarking on Wallis's related use of interpolation tables to derive the series for the quadrature of the circle, Whiteside observed:

> this reasoning by analogy—or perhaps more correctly from a feeling for a general pattern which seems to run through a set of particular results—exemplifies a process which must be fundamental to any system of interpolation: since there are an infinite number of ways of filling in a pattern, we choose that way which seems best suited (in a sense wider than the strictly mathematical), best conforms, to the instances known. Wallis' assumptions in his derivation are quite audacious, and in a rigorous treatment must be carefully justified—yet in following through an intuition that he was thereby achieving a result which is both true and important Wallis was doing something practiced by every creative mathematician, however lucky in that he did not seriously have to consider the boundary-cases where such general reasoning by pattern must break down. (Whiteside 1961, 240–241)

Wallis took a few concrete results to indicate the validity of a general approach to the study of certain properties of curves. He had no proof of the approach's validity. What he *did* have was a way of setting up these problems that was obviously highly adaptable to different problems that had traditionally been approached geometrically. The motivation for adopting that problem set-up lay in his illustration of its limited but tantalizing success in a few different contexts, with the clear indication that it could be further generalized and extended by other mathematicians to "find the like." These successes would have drawn a crowd, to use Brush's terms. But only the promise of further extension could have sold the method itself. Where classically oriented mathematicians like Fermat, Hobbes, and Huygens were discomfited by the departure from the demonstrative rigor of geometrical proof, a new generation of mathematicians—mathematicians like Newton and Mercator—saw the infinite possibilities promised by such an approach.

The world does not tell us what is analogous to what, or what is the "correct" way to project beyond our evidence. As Wallis pointed out in response to criticisms of his use of "induction," this was the same method "used by Euclid every time he allowed a triangle to stand for any other triangle of the same kind" (Stedall 2005, 29). The choice to infer of all right triangles what we can prove of only a few, no matter how intuitive, is still a choice; we are required neither by nature nor by powers higher than that to treat the possession of a right angle as constitutive of membership in a class. Somehow, the ability of a given right triangle to function as an exemplar is coincident with our certification of that class.

Wallis's infinite series were able to function as exemplars not because they solved outstanding problems, nor because they convincingly succeeded in delivering correct results. They did neither; Wallis could offer nothing in the way of a traditional demonstrative guarantee that, for example, a series for any arbitrary number of lines n under a parabola could be reduced to the sum of $1/3 + 1/6n$. Rather for those willing to treat them as arithmetic representations of traditional geometric problems, his infinite series functioned as a sample of a new form of problem set-up, a new approach to problem solving. It was this *willingness* to treat Wallis's approach as a form of problem-solving that marked the accelerated pace of the development of mathematics after 1656. Newton was not unaware of the conceptual problems that plagued infinitesimals; an entry in his *Quaestiones quaedam Philosophicae* (probably from 1664) is a succinct articulation of the essential paradox to which Hobbes alludes above (MS Add. 3996: 89r; see Newton and Whiteside 1967, 89–90). Newton chose not to let it bother him. His contemporaries made the same choice. Regarding this choice, Morris Kline (1972) wrote: "The general lack of concern . . . is surprising, in view of the

fact that the Europeans were now fully aware of what rigorous deductive mathematics called for."

> Nevertheless, the mathematicians proceeded blithely and con-
> fidently to employ the new algebra. . . . Without realizing it the
> mathematicians were about to enter a new era in which induc-
> tion, intuition, trial and error, and physical arguments were to
> be the bases for proof. The problem of building a logical foun-
> dation for the number system and algebra was a difficult one,
> far more difficult than any seventeenth-century mathematician
> could possibly have appreciated. And it is fortunate that the
> mathematicians were so credulous and even naive, rather than
> logically scrupulous. For free creation must precede formaliza-
> tion and logical foundations, and the greatest period of mathe-
> matical creativity was already under way. (Kline 1972, 282)

The promise of Wallis's approach—its exemplification of a new
kind of problem set-up—was evident enough to attract them away
from competing modes of mathematical activity.

In the remainder of this chapter, I want to understand in detail
the process by which exemplars emerge and come to form the basis
for further inquiry. I begin with the warranted presupposition that
groups of practitioners often come to exhibit a shared sense of how
scientific inquiry ought to be conducted, broadly speaking. Or they
transition from one shared sense to another. In either case, there
emerges a new set of normative constraints on what practitioners
recognize as an effective way to structure inquiry—an approach to
understanding nature or mathematics that will tell us more than
we now know, and more than any other known approach is likely
to. I make the further warranted presupposition that practitioners'
shared sense of the way that inquiry *ought to be* conducted is a
descendent of some concrete instance of how inquiry *has been*
conducted. These concrete instances include such epic events as
Kuhn's achievements "sufficiently unprecedented"—*Arithmetica*

Infinitorum, the *Origin*, the *Principia*, and other great works. But they also, probably far more frequently, come in the form of much smaller-scale innovations, innovations whose domain of applicability might not reach beyond a narrow subdiscipline.

We now want to know how an instance of research comes to be regarded as representing a new class—how the transition from instance to "type-specimen" occurs. Are there certain features of an investigation that are particularly facilitative of a researcher's willingness to treat it—or aspects of it—as denoting an entire class? Are there certain specifiable research conditions under which researchers are more inclined to regard an attempt to solve a problem as a *sample* of problem solving?

Second, we want to know how the transition from *is* to *ought* occurs: how do normative constraints on the practice of science emerge from concrete instances of scientific practice? More specifically, we want to know the process by which practitioners come to view an exercise in inquiry as embodying information about how and what they themselves *should* investigate. Under what conditions are practitioners willing to treat an instance of inquiry as more than just another brick in the wall, or a theoretical exercise— that is, under what conditions are they apt to view an instance of research as possessing normative significance? Throughout our search for an answer, let us keep in mind the immense weight with which the question "On what shall I work?" consistently bears down on the practitioner.

Lastly, we should like to know *why* these two phenomena—(1) the attribution of normative significance to a problem-solving attempt and (2) the elevation of a problem-solving attempt to the status of a *sample*—characteristically occur together in the context of scientific inquiry. Are they two essential aspects of a single process? Is it possible for a researcher to attribute normative significance to a problem-solving attempt without elevating it to the status of a sample? Conversely, are there conditions under which a researcher regards a problem-solving attempt as a sample of problem solving, without endowing it with normative import? To show that the two phenomena are conceptually linked would be a profound result indeed.

These two sets of questions are philosophically important because they represent the pathway by which a new conception of scientific rationality might be articulated, one which satisfies the demand for a philosophical explanation of how decisions regarding which problem-solving approach should guide future research can be epistemically warranted without appealing to a track record of problem-solving successes. This demand arose out of the failure of Kuhn and others to provide an adequate epistemology of the transition from one problem-solving framework to another. Kuhn's proposals—an earlier faith-based one and a later semantic one—are failures in that they give no account of why scientific inquiry has been so successful. It is not even that they give no account. The distinctive success of scientific inquiry makes no sense whether we accept Kuhn's early claim that such transitions are not governed by reason, or his later claim that they are governed by reason as a matter of definition. But we learn something from these failures. Specifically, we learn that whatever account we end up giving will need to clarify (1) why the norms of decision governing inquiry are causally sufficient for producing the pattern of problem-solving success evinced by the history of scientific inquiry; (2) how those norms of decision correspond to a philosophically defensible understanding of rational choice in the context of scientific inquiry.

4.3. Exemplification

Following Kuhn, I have accepted that works like the *Origin* or the *Principia* served as exemplars for entire communities of practitioners. Kuhn never says how; the most specific thing he says is that practitioners model solutions to new puzzles from previous solutions. That's true as far as it goes, but its inadequacies are legion, and they are unfortunately part of Kuhn's legacy in many circles. I introduced the first major inadequacy in the previous chapter—viz., it ignores the historical fact that practitioners model their solutions to new puzzles on unsuccessful attempts and on outright failures just as routinely as they do on solutions (or what count as solutions by their lights). The preceding vignette from seventeenth-century mathematics is one among countless illustrations of this historical fact. The second major inadequacy is that the lack of specificity occludes any potential resources we might find for understanding the rationality of paradigm choice in the nature of the exemplification process. This is how Kuhn wanted things to sit: if exemplification is a matter of "base perception"—as when Galileo "learned to see" a pendulum as a kind of inclined plane—then there essentially are no underlying details that can serve the purposes of reconstructing a process of rational deliberation which eventuates in the embrace of a particular research framework. However, as we will see later, the resources *are* there, and they're much easier to see when we've discarded the notion that exemplars emerge only from solved problems.

I want to begin our examination of this process by considerably refining the notion of exemplification as it has been used thus far. The route to refinement is going to proceed through a few stages. I first want to run through a few clear-cut "cases" as a way of showing why Nelson Goodman's classic account of exemplification works as well as it does. It ably captures paradigm cases of exemplification; it ably excludes cases which display a failure to exemplify. I'll then look at a few less-than-paradigmatic cases, where Goodman's gloss also succeeds.

4.3.1. An Exemplary Goiter

During her rounds one morning, Dr. Layla, a leading endocrinologist with decades of clinical experience, examines my goiter. Struck by its perfectly characteristic manifestation, she decides to bring all the endocrinology residents in the room to gawk at the goiter. As they huddle around me, Dr. Layla emphasizes certain aspects of my goiter's shape and position by pointing at different regions on my throat and describing features to which she is attempting to draw the students' attention.

Goodman argues that exemplification is "possession plus reference," which nicely explains what makes my goiter an exemplar: (1) it possesses the property of being a goiter; (2) Dr. Layla means for that goiter to symbolize goiters—to refer to the predicate 'goiter'—and her students take the goiter as such. Suppose that, following the clinical exam, Dr. Layla then pointed to a diagram on the wall depicting a view of a goiter from inside the neck. Again, Goodman's account does an effective job of distinguishing between this diagram and goiter exemplification. As in the exam, the diagram successfully refers to the predicate 'goiter.' But it fails to exemplify goiters because it is not itself a goiter; it does not fall within the extension of the predicate 'goiter.'

Goodman held that possessing a property was necessary for exemplifying it, but not sufficient. This becomes obvious once we focus in on some of my goiter's more "Cambridgey" properties. Even though my goiter has the property of being in Cleveland, it does not exemplify being in Cleveland. On Goodman's account, what my goiter lacks as an exemplar of the class 'things in Cleveland" is that my goiter does not refer to that class. For him, exemplification is possession plus reference. My goiter *can* exemplify any class it falls into, but it *ends up* exemplifying only those classes to which it refers.

A fair question to raise at this point is, why doesn't my goiter refer to the class of things in Cleveland? Goodman gives some rough and ready guidance on this issue—he says, reasonably, that, as with all referential capacities, whether a symbol refers to a class "depends upon what particular system of symbolization is in effect" (Goodman 1968, 53). The fact that the inscription, "tiger," refers to tigers is not a power that is intrinsic to that inscription; it is an artifact of the symbolization system that characterizes the English language. Where my goiter is concerned, it is easy to motivate the claim that it does not exemplify being in Cleveland, because it is not credible to assert that the symbolization system which prevails in that particular clinical setting is one in which my goiter refers to the predicate, 'being in Cleveland.'

But that there might in principle be such a system of symbolization is conceivable. One day, Dr. Layla announces to her residents that she's going to take them to see a number of patients with problems that are characteristic manifestations of a range of endocrinological marvels that are unique to the Cleveland area. After visiting a couple dozen patients, the residents eagerly pour into my suite to gaze at a classic case of "Cleveland goiter"—my goiter's tendency to expand and contract throughout the day as a consequence of the very particular way in which my iodine deficiency interacts with the astonishingly high levels of estrogen found in Cleveland's drinking water.

How does my goiter exemplify being in Cleveland in the second example but not the first? Here is where the vagueness of Goodman's appeal to "symbolization systems" comes in handy. The system of symbolization in operation is not merely what can be uttered or inscribed, but includes anything that might function as a vehicle for content in a given context. In this case, the residents have been instructed ahead of time that the conditions they'll be observing are characteristic of being in Cleveland. This instruction affects the symbolization system in effect. By emphasizing that the reason why they are visiting this particular set of patients is because each of them has an endocrine problem that only arises in Cleveland, Dr. Layla has made salient the property of being in Cleveland. Thus, the symbolization system that subsequently prevails during these patient visits will afford prominence to that property. By contrast, had Dr. Layla made no prior mention of the overt Clevelandishness of the patients' problems, the symbolization system would not have granted particular salience to the property of being in Cleveland, and my goiter would not have referred to it. By making the property of being in Cleveland salient in this context, Dr. Layla has changed the prevailing symbolization system in a way that ensures that my goiter will refer to that property.

This example illustrates two central features of the nature of exemplification that will figure prominently in the next couple of chapters as our understanding of exemplification in science further develops, features which underscore the inadequacies of Kuhn's subtle whisperings on the subject. The first feature is the extreme context sensitivity of reference. That the same object might equally well exemplify one property in one context and another in some other context is an expected consequence of the manner in which referential relations are prone to shift given even very slight variations in conditions. Thus, to say that practitioners use previous problem-solutions as exemplars leaves a lot unsaid about why those previous solutions tend to be uniformly understood by members of a research community as referring to the same property. The manner in which seventeenth-century mathematicians followed up on the work of Cavalieri and Wallis exhibits a degree of uniformity that should surprise us, given the extreme context sensitivity of reference. So there's a puzzle there: *exemplars seem to possess something that buffers against this context sensitivity.* Second, the "Cleveland goiter" example shows the way in which speakers and hearers are able to exert direct control over certain referential capacities; in Goodman's words, "I can let anything denote red things" (Goodman 1968, 58). When Dr. Layla wanted my goiter to refer to the property of being in Cleveland, she constructed a context in which that property of my goiter was particularly salient. In this connection, Elgin observes that "[o]ne of the powers of exemplification as a referential device is that we can normally improvise exemplars at will" (Elgin 2017, 188). Again, this contrasts with Kuhn's picture, which goes out of its way to depict exemplification as outside the conscious control of the practitioner.

4.3.2. The Exemplification of Scientific Problems

At the end of the last chapter, I argued for the claim that promise inheres in the evidence provided by problem structuring, rather than problem solving. Using Kuhn's appeal to the promise perceived within exemplary works of science as a starting point, I pushed us toward a significantly more liberal view of exemplars that is characterized by a few distinct qualities. First, I dispensed with the notion that exemplars emerge from solved problems. Second, I suggested that what exemplars exemplify is a *kind* of problem, characterized by a certain "structure." Third, I argued that a structure is something that practitioners impose on a problem, often consciously and specifically with problem solving in mind. Lastly, I claimed that whether two problems belong to the same kind comes down to whether *practitioners are willing to structure them* in a similar way.

Let's look at how Goodman's analysis of exemplification—"possession plus reference"—allows us to organize these features into a coherent picture of the process of exemplification in organized inquiry. Beginning with the "possession" requirement: if what a scientific problem-solving attempt exemplifies is a certain structure, then any attempt would have to possess said structure to qualify as an exemplar. It's easy to see how this works in the context of our discussion of $f = ma$ from the previous chapter. $f = ma$ functions as an exemplar partly by embodying a specific quantitative relationship between force, mass, and acceleration. That specific quantitative relationship is the "structure" it possesses; it is a "schema," or an organizing principle, or whichever cognate notion you like—something that instantiates a pattern that can be reproduced in other contexts.

The sense in which Wallis's series "possess a structure" is, I think, much less straightforward. Let us assume *arguendo* that one of the primary contributions of *Arithmetica Infinitorum* was to show that some problems of area and volume could be structured as the arithmetic sum of an infinite series of terms. Is there anything in the book that "possesses" the structure of an infinite series? It seems to me that the most promising candidate is Wallis's use of a very limited series of demonstrations followed by the symbol "&c", meant to suggest that the pattern continues to infinity. Apparently, this was enough for Wallis and for his readers. It is, for example, the same form taken by Newton's arithmetic quadrature of the hyperbola; Figure 4.3 contains no fewer than five series followed by "&c". His expression for the general binomial theorem also follows this pattern (Fig. 4.4). And it is the same pattern Mercator uses to express the infinitely declining magnitude of the line segment FH in his arithmetic quadrature of the hyperbola (see Fig. 4.11), among a variety of other places in Logarithmotechnia. Regardless of what we might think about it, this pattern was clearly mathematically coherent enough for many of seventeenth-century Europe's best mathematicians to use as a method for organizing their investigations into problems unaddressed by Wallis. What might perhaps be a "porto-structure" for us was, for them, a bona fide way of setting up problems in mathematics. Viewed from this perspective, Wallis's examples possess the structure of an infinite series in the sense that, for him and his readers, it simply made sense to them that the series should continue to exhibit that pattern ad infinitum. Of Newton's generalization of the binomial expansion, Morris Kline wrote: "he *became convinced* that the expansion held for fractional and negative *n* (it is an infinite series in this case) and so stated, but never proved, this generalization" (Kline 1972, 273; emphasis added).

Atque ita continuatâ operatione, deprehenditur $\frac{1}{1+a} = 1 - a + aa - a^3 + a^4$ (&c.) $= FH.$

Figure 4.11 Mercator's series for the hyperbola.

Thus, the question of whether a problem has a certain structure, like the question of whether an object possesses a property more generally, turns out to be more complicated than a simple "yes" or "no." For some mathematicians—*good ones*—Wallis's examples had a discernible and reproducible structure to them. His critics, insofar as they did not find his arguments mathematically compelling, denied that there was any reason to believe that the pattern exploited by Wallis disclosed the existence of infinite series. Where some mathematicians perceived a structure, others sensed a mirage. What this suggests is that a problem's possession of a structure has more to do with a practitioner's *willingness to attribute that structure to the problem* than it does anything else.

There is nothing special about structure per se that makes a problem's possession of it contingent on a willingness to attribute; lots of properties are like this. For most of us, an explosion is not a collision. But, as I remarked in Chapter 2, there is a subset of special people for whom explosions *do* possess that property—or, at least, that is how they are represented in the context of problem solving. These people are members of a linguistic community in which the prevailing symbolization system assigns the property of being a collision to each explosion. I can see why, if you were a member of that community, you'd be motivated to frame explosions in that way. You do a lot of calculations, and it makes your job easier. I get it. I'm mercifully unburdened by any such membership in that I am not now, nor have I ever been, a physicist. To me, it is beyond unintuitive that an explosion has the property of being a collision. Frankly, it's perverse. Based on the way I and the rest of the statistically normal people use that term, nothing could be less collision-y than an explosion. I don't think there's any perspective-independent fact of the matter as to whether or not an explosion has the property of being a collision. If you find yourself inclined (for admittedly defensible reasons) to partake of a symbolization system that frames explosions as collisions, I support you. Just don't hold it against the rest of us if we can't bring ourselves to assign that property to an explosion, okay?

This brings us to Goodman's second criterion: the denotation of a class. In the previous chapter I asserted that what exemplars in science exemplify is a *kind of problem*—a problem-kind. A problem-kind is characterized by a certain structure. So, for example, problems in dynamics form a problem-kind that is characterized by the $f = ma$ structure. Wallis's examples form a problem-kind that is characterized by structuring problems of area and volume as arithmetic sums of infinite series. &c. In what sense do, say, Wallis's examples *refer* to a *kind* of problem characterized by that structure? In a manner similar to property attribution, the denotation of a class of problems occurs at the moment a problem-solving attempt *is treated as a sample* of how to structure problems to make them solvable. This willingness to treat problem-solving attempts as such is, in effect, the inauguration of a new symbolization system, one in which Wallis's attempts *stand for* the kind of set-up they themselves possess. Their status as symbol is imposed on them by practitioners.

4.3.3. The Salience of Set-Up

An exemplar is an individual member of a class for which it is also a symbol, a class wherein membership is defined by possession of a certain cluster of properties. The exemplar, being a member of the class, possesses this cluster of properties. It is able to *function* as a symbol for that class by being *treated* as a symbol for that class. Our reasons for treating it as such are our own.

The variety of exemplar involved in what Kuhn called paradigm choice is an individual problem-solving attempt in possession of a discernible structure or "set-up" (Kuhn's term). We've discussed at length Wallis's novel set-up in *Arithmetica Infinitorum*, which is not an example used by Kuhn, nor could it have been; it violates Kuhn's "solved problem" condition. However, each of Kuhn's examples in

Structure can be redescribed as exemplary of a kind of problem set-up, rather than a solution. Since this is at least partly not a book about *Structure*, you'll just have to trust me that this is possible—which I don't think is asking a lot, really. That claim is *prima facie* much more plausible than Kuhn's own perspective. Think about it: even if each of his cited exemplary works of science unambiguously produced solutions to problems—not just convincing accounts of phenomena, but genuinely cosmic revelations about Reality—practitioners might easily not have any means by which to advance inquiry. The promise of further inquiry is not built into the concept of truth. And truth's facilitation of further inquiry has for far too long been taken for granted. In any event, at the very least we now have some semblance of an explanation for how exemplars are generative of future inquiry—they provide a sample of how to set up problems in a way that makes them solvable.

What we do not yet have is an explanation for how a given problem-solving attempt comes to acquire its amazing referential powers. Why, for example, were Wallis's examples taken as samples of how to set up problems of area and volume, rather than, say, samples of what can be done with numerals in a book otherwise composed in Latin? Because that is also a property that they possess. Why are they treated as samples at all? This familiar philosophical gambit generalizes. For instance, what is it that endows my goiter with the power to refer to the class "Cleveland goiter"? Notice that this is not something of which it is routinely capable. As I walk down the street in Cleveland, my goiter fully exposed, passersby are not struck by how exemplary it is of Cleveland goiter. They're mostly thinking, "What is that?" Most of them wouldn't know a goiter if it punched them in the face, let alone possess the ability to distinguish an instance of Cleveland goiter from some garden-variety goiter. Obviously they can't be expected to appreciate the awe-inspiring degree to which my goiter possesses the characteristic properties of Cleveland goiter. Only the most discerning goiter enthusiasts are blessed with that capacity—enthusiasts like a certain Dr. Layla. Her cultivated sensibility for goiter classification is something that, for example, her students do not share. When she coaxes them into the room to gaze upon my goiter, they don't realize that they are standing in the presence of goiter gold. They have no idea what it even is. She begins by informing them that it is a goiter, and then proceeds to emphasize certain features, drawing arrows and circles on my neck with special ink that you can wipe off of skin. As my goiter's Clevelandish-ness kicks in, she draws their attention to its radical change of shape and position, noting that goiters outside of Cleveland rarely behave that way.

My goiter's power to exemplify in this instance harkens back to our earlier observation regarding the infamous context sensitivity of reference. Here we see an example of how the system of reference can be manipulated to endow an object with the powers of exemplification. Through her targeted accentuation of the salience of certain features of my goiter, Dr. Layla shifts the semantic context in which my goiter is interpreted. In doing so, she has elevated its status from a humble symptom of my iodine deficiency to a monument to all of goiter-kind. Her repeated emphasis of certain features, her efforts to get them to stand out from the background, her explicit references to their contrast with other goiters—all of this contributes to the increased salience of those features in particular. Her intent to use my goiter as an exemplar of "Cleveland goiter" depends on these efforts. Both she and the students understand that they are to treat my goiter as a sample of *something*—this is one of the interesting background presuppositions of the instructional context, and one of the things that distinguishes it from a "walk-down-the-street" context. Dr. Layla has engineered a system of reference in which my goiter is supposed to stand for the class "Cleveland goiter," and she has accentuated the salience of certain features to allow my goiter to stand for that class rather than another class—say, the class of things that can change shape and position, or the class of things on my neck. By explicitly instructing the students to treat my goiter as a sample of Cleveland goiter, and by accentuating the salience of these particular features as symptomatic of Cleveland goiter, she is attempting to impart to them a disposition to characterize certain other growths they might encounter as members of the class "Cleveland goiter."

Can considerations of salience explain why Wallis's series were similarly imbued with the power to refer to, and thus the ability to exemplify, a new kind of problem set-up? I believe so. The key lies in his *repetition* of the infinite series set-up across a range of different problems. Proceeding through *Arithmetica Infinitorum*, what one sees over and over again is the reformulation of a classic problem in geometry using an arithmetic sequence of terms that get added together to *indicate* (i.e., not fully prove) a result previously known to be correct. "The above and similar examples were enough to convince Wallis that the method worked, and he set out to extend it by investigating sums of cubes and higher powers" (Stedall 2005, 27). By repeatedly displaying the infinite series set-up across a range of different problems, Wallis amplified the salience of that set-up to the point where it emerges as something like a general theme of the book. As he says, "the design of that Treatise . . . was . . . to shew a way of Investigation or finding out of things yet unknown . . . and how others might (by those Methods) find the like." This goal was accomplished through repetition.

The repetition of a feature is one of a handful of general ways to amplify its salience. We saw another in the curious case of Dr. Layla and the goiter: *emphasis*. And in *Structure*: *solved problems*. In addition to these domain-general salience amplifiers, each discipline has features that stand out to practitioners trained in a given tradition. Lange (2014) mentions symmetry and unity as features that make a result salient for mathematicians. As Kuhn would suggest in the years after *Structure*, the purpose of certain model problems or systems in the training of practitioners is to impart to them an understanding of what matters for solving problems in that discipline. Those model problems and systems accentuate the salience of certain features that have been found to reward a greater degree of focus relative to the rest of experience. The rise of classical mechanics, for example, owed much of its power to the discovery that focusing on *acceleration* rather than *velocity* facilitated a far deeper and broader understanding of motion than had previously been available (Westfall 1977, chapter 1). From the early seventeenth century onward, the problem of free fall was used as something like a conceptual laboratory for every generation of practitioners from Galileo to Newton because of how the *invariance* of acceleration across objects in free fall accentuates the salience of acceleration in this context (Damerow et al. 1991).

We began by promising to address Nickles's complaint that Kuhn never explains the seemingly magical transition from solved problem to exemplar. That promise has been fulfilled, except that I used problem structures instead of problem solutions as the thing exemplified. Specifically, I tried to show how can we trace that magical transition through the lens of Goodman's account of exemplification: "possession plus reference." The "reference" condition that Goodman places on exemplification adds a lot to the original Kuhnian picture, but it does not give us everything, because it does not give us a fine-grained conception of how that reference condition comes to be satisfied. It would be beyond both the scope of this book and of my own understanding to attempt to articulate a more detailed picture of the causal process by which an object comes to refer to a class wherein membership is defined by that object's most salient properties. I'm not sure how much it would add to the specific project we've taken up in this chapter, however. We know that relations of reference are acutely sensitive to changes in context, changes such as which properties stand out as particularly salient. It also seems more than probable that the amplification of a property's salience contributes to its role in exemplification. The "goiter" example illustrates how amplifying the salience of certain properties can shift the system of reference to one that treats those properties as diagnostic of a class or a kind. As I have argued, the "Wallis" episode is meant to show a real context in which (1) the salience of certain features was amplified, and (2) practitioners wound up treating the objects that possessed those features as samples of a kind—in this case, a kind of problem. The relevant features in this context were the arithmetic sequences contained in *Arithmetica Infinitorum*, made salient through repetition, and eventuating in the emergence of a new schema for setting up traditional problems of area and volume.

The history of infinite series, and of mathematics in general, is a convenient way of illustrating the sense in which exemplars exemplify a kind of problem set-up or structure. Is it *too* convenient? Does the highly formalized nature of problem set-up in mathematics lend itself too easily to the idea that what matters in historical episodes of exemplification is the emergence of a certain general strategy for structuring problems to make them solvable? After all, Kuhn's exemplary great works of science run the gamut from the formidably formal (*Almagest*, *Principia*), to the respectably formal (*Opticks*, *Chemistry*), to the relatively informal (*Electricity*, *Geology*). Since we are accepting *arguendo* that these works did historically play the role Kuhn attributes to them, contributions like these ought to exemplify a kind of problem set-up as well.

I've always found it a bit curious that Kuhn did not include Darwin's *Origin* in that original list. As a paradigm-inducing achievement, only the *Origin* occupies the same stratum as *Almagest* and the *Principia*; the effects of the other four works were not nearly so pronounced as these. At the same time, the *Origin* is probably the least "formal" of the entire lot. There are no equations. There are relatively few reported measurements; and the scope of those tends to be quite narrow, as for example when Darwin reports on how many days a seed could soak in sea water and still germinate (answer: a surprisingly large number of days). The *Origin* is as much a beautifully written work of literature as it is a monumental contribution to science, which is not something that could be said of any of the other works. Its many anti-formal virtues—combined with it having unambiguously solved a problem—make it an important test case for the claim that it exemplifies a kind of structure or set-up that practitioners could use to make problems of natural history solvable.

Philip Kitcher (1985) argues persuasively that Darwin's achievement was the establishment of "schemata" or "a collection of problem solving patterns" for answering "general families of questions" about organisms:

> What Darwin constantly emphasized, and what his contemporaries recognized, was that the *Origin* was not only a confession of ignorance but also a *structuring of our ignorance*. . . . [I]ts primary accomplishment lay in identifying the questions that biologists ought to ask. It is because of this primary accomplishment that Darwin may truly be said to have revolutionized the field. The nature of that revolution is captured in one of Hooker's letter to Darwin:
>
>> "But, oh Lord!, how little do we know and have known to be so advanced in knowledge by one theory. If we thought ourselves knowing dogs before you revealed Natural Selection, what d— d ignorant ones we must surely be now we do know that law."
>
> (Kitcher 1985, 60)

In this way, Darwin's theory "claims for scientific investigation questions which rival theories dismiss as unanswerable" (Kitcher 1985, 51).

This episode is well-captured by the model we've been developing. The volume of empirical knowledge of natural history possessed during Darwin's time was substantial. Indeed, it had become substantial so rapidly that, by the beginning of the nineteenth century, the community of naturalists was in a state of profound confusion (Ruse 1979). In no sense is this better illustrated than in the unwieldy set of general theories that proliferated during that time (Rehbock 1983; Corsi 2005); Darwin reports his own despair of "a theory by which to work" at this time in his *Autobiography* (Kohn 1980, 100). The *Origin* forged a way out of this quagmire by providing "schemata" for organizing the abundance of knowledge, highlighting the importance of certain kinds of information—information about biogeography, ecology, embryology, genealogical relatedness, and geology—that contribute to a complete causal understanding of natural history. In Kitcher's words, Darwin's schemata advance inquiry by providing a structure to "general, unfocused, explanation-seeking questions about organisms" (Kitcher 1985, 50). This structure succeeds by breaking down previously unanswerable questions into distinct, investigable components, applicable to all aspects of natural history. Solving the problems of adaptation and species diversity drew a crowd. But what "attracted an enduring group of adherents away from competing modes of scientific activity" was the problem-solving framework or set of schemata exemplified by Darwin's arguments in the *Origin*.

The now-familiar signatures of exemplification can be seen by looking at the argument for the principle of natural selection that develops over the course of chapters 1–4 of the *Origin*. As Kitcher and others have emphasized, this argument does not employ any novel information. Indeed, its principal claims—that organisms vary, that their children sometimes inherit those variations, and that some variations are more advantageous than others—had been commonsense for ages; ages and ages. Darwin's particular emphasis on these platitudes, and his repeated illustrations of them, allowed them to stand out against the centuries-old backdrop of which that had long been components. The subsequent argument for descent with modification proceeds in like fashion, albeit with empirical premises generally of a more recent vintage. By repeatedly using the same varieties of information (e.g., biogeographical information, embryological information, etc.) across a range of explanatory contexts, the *Origin* exemplified a general schema for resolving questions about natural history. As the salience of these facts comes into sharper and sharper relief, they transition from mere reportage into illustrations of *how to structure* certain kinds of investigations. Indeed, what else *could* they be? Darwin's argument regarding the modification of the shape of honeycomb cells in chapter 6, for example—what is its purpose? It is, as Brandon (1990) argues, a "how-possibly" explanation for the different shapes of cells made by different species of honeybee. These species are part of a lineage, and the different shapes mark slight modifications of the responses to the problem of storing honey that persists across the lineage's history. That *might* be how things happened, which is all the argument is equipped to show (Lustig 2008, 109–112). Empirical investigation is required to see whether Darwin's account can actually be borne out. But the argument has a certain structure to it, a structure that he uses repeatedly to deal with various "difficulties on theory" and to render his general view of descent with modification as plausible as possible. One need not accept any of Darwin's specific claims about the descent of *this* from *that* or the modification of *this* for *these* purposes to see how *some* version of such an argument might be supported empirically, and thereby rendered convincing to other naturalists. As Darwin appeals to this structure over and over again, an abstract pattern emerges in the mind of the careful reader in desperate search of "a theory by which to work."

Conclusion

Certain works of science and mathematics "serve for a time implicitly to define the legitimate problems and methods of a research field for succeeding generations of practitioners." This is to say that they are exemplars. They legitimize problems and methods by constructing a system of reference in which the use of a particular method or the set-up of a particular problem functions as a sample of how to structure a problem so as to make it solvable. This system of reference is constructed out of new categories that emerge through the amplification of the salience of certain properties, properties which come to define the category itself. In the context of formal inquiry, the relevant category is a kind of research problem, characterized by a certain "structure," "schema," or "frame"—a set of ordering principles that encapsulate how to generate solvable problems in a given discipline. It is to these structures/schemata/frames that I turn in the next chapter.

This is the full-throated alternative to Kuhn's underspecified and inaccurate account of exemplification in rational inquiry. Ultimately, the process I've described must be sufficient to bear the weight of promise and its warrant as described in the previous chapters. The remainder of the book is devoted to meeting this challenge.

5

Framing Fruitfulness

Copernicus, ignorant of his own riches, ever took it upon himself to express Ptolemy, not the nature of things, to which, nonetheless, he of all men came closest.
—Johannes Kepler, *Astronomia Nova*, chapter 14[1]

5.1. The Normative Significance of My Goiter

So my goiter is an exemplar. Of course it is. Dr. Layla made sure of that. Still, a lot of people might ask, "But, Chris, what's the *normative significance* of your goiter?" Does my goiter, for example, "define legitimate problems and methods," as some exemplars are reported to do?

There's an issue of scale here that needs to be clarified. Obviously my goiter is not an exemplar on the scale of the *Principia* or the *Origin*. No one is saying that. It's not equipped to define an entire branch of modern science. It's probably not even fit for a cover article in *GQ* (*Goiter's Quarterly*). No, my goiter's capacity to exercise a normative influence on inquiry is most likely limited to those precious few students lucky enough to have struck "goiter gold" during their endocrinology rotations.

What is that influence, and what is its source? As to the first question, let's think about what Dr. Layla's students have acquired through their little training session. The purpose of that session,

[1] Thank you to Aviva Rothman for this reference.

Fruitfulness. Chris Haufe, Oxford University Press. © Oxford University Press 2024.
DOI: 10.1093/oso/9780197666395.003.0005

recall, was not merely to introduce students to my specific goiter, but to impart to them a disposition to classify growths with certain features as instances of "Cleveland goiter." Because my specific case of Cleveland goiter is so "classic"—that is, because it possesses the defining features of Cleveland goiter in their purest form—and because Dr. Layla has gone about amplifying the salience of those features yet further still, the students end up acquiring the means with which to diagnose other growths as instances of Cleveland goiter.

These acts of classification have normative consequences. Indeed, it is *intended* that they have normative consequences. Cleveland goiter is a problem that needs to be solved, and there is a prescribed course of action which attends the act of classifying a growth as another fascinating case of Cleveland goiter, a problem-solving protocol that is expected to result in the solution to that problem. Thus, by acquainting her students with my exemplary goiter, Dr. Layla has effectively imparted to them knowledge of the conditions under which it is appropriate to avail themselves of a certain approach to problem solving. So, yes, my goiter is normatively significant. Like other exemplars, it serves as the basis for prescribing an approach to solving a problem. To classify a growth as an instance of Cleveland goiter is to prescribe a specific approach to the problem that makes the problem solvable.

In this chapter I look in more detail at the ways in which exemplars like Wallis's *Arithmetica Infinitorum*, Newton's *Principia*, and, to a somewhat lesser extent, my goiter, come to define legitimate problems and methods for a research field. I have been discussing the notion of classes of problems or problem-kinds, which are marked off by a certain way of structuring or setting up problems. In the account I've been developing, these classes or kinds emerge from exemplars as abstractions of the problem set-up or structure that those exemplars exemplify. At that point, they begin to function as a general template for approaching specific types of inquiry.

Kuhn called these exemplified structures "paradigms." In a broader context, Nelson Goodman (1968) called them "schemas"; Elisabeth Camp (2019a, 2019b) calls them "frames." At the core of this polyonymous notion is a set of ordering principles for a given domain, principles that tell us what matters for solving a problem, and how the things that matter relate to one another. The ordering principles might dictate a highly formalized structure, as in those that emerge out of Wallis's study of infinite series. Or they might recommend something far more qualitative, as in the scientific framework that arose following Darwin's *Origin*. The formal specificity of the ordering principles is not the point. The point is that there *is some set of ordering principles* through which practitioners can express problems that they can work on. As Kuhn argues persuasively in *Structure*, these principles are rarely articulated. They are exemplified, and they are essentially absorbed in the course of professional training and practice.

It is instructive to contrast this order-centered notion of *paradigm* with one articulated in Kuhn's *Structure*. Earlier I claimed that problem set-up receives no explicit attention in *Structure*, and that Kuhn's appreciation for it does not appear until "Postscript." But what *does* receive explicit attention in *Structure*? Kuhn's focus there is very much on phenomena themselves; a paradigm shift, a "change in world view," is a transformation of what practitioners deem to be the sort of

> facts . . . [that are] . . . particularly revealing of the nature of things. By employing them in solving problems, the paradigm has made them worth determining both with more precision and in a larger variety of situations. (Kuhn 1962/2012, 25)

Maybe what Kuhn came to realize in the years leading up to the refinement of his views in "Postscript" was that, even if some phenomena are as revealing as a tell-all autobiography, they are not going to be pressed into the employ of problem solving unless there is a way of structuring problems in which those phenomena feature and which produces solvable problems. The shocking adaptedness of organisms was not something that we required the *Origin* to disclose. But it was not until the *Origin* that we were able to, as Kitcher stated, fit adaptedness into a problem-solving schema that made questions about it scientifically answerable. Similarly, it took no prodding or change in perception to convince particle physicists during the 1960s that it might be revealing to study the short-distance interactions between particles. However, they lacked a set-up for doing the calculations, and until they got one, those facts made no meaningful appearances in problem-solving efforts.[2] To put it in terms that *Structure*-Kuhn might have understood: no ordering principles, no normal science.

In fact, something like the reverse of the process imagined by Kuhn often seems to be operating. That is, it is often the birth and subsequent refinement of a problem-solving structure that amplifies the salience of certain phenomena and drives the growth in our perception of their importance for achieving understanding. Both Galileo and Descartes emphasized the special status of inertial motion. While Galileo's conception is a hybridization of his novel views and his Aristotelian baggage (Westfall 1971, chapter 2), Descartes's is what we would recognize as the "classic" version inherited by Newton. But it was the Newtonian problem-solving schema that catapulted inertia to its vaunted status as a central feature of our understanding of motion (Westfall 1977, chapter 1). The set of equations known as the Lorentz transformations reveal inertia to be an even more puzzling property of motion than the founders of classical mechanics could have imagined (Einstein 1920; Brown 2005).

---

[2] My friend, colleague, and particle physics storyteller, Cyrus Taylor, recalled a time during the early 1980s when he walked in on his PhD advisor in the act of throwing out a bunch of journal issues. "What are you doing?" he asked. "I'm throwing out copies of *Physical Review* from the 1960s." When Cyrus asked why, his advisor responded, "Because nothing happened." (Remember, no journal articles were online at this time. He wasn't getting those back in a more convenient form.)

Together, these contrasts with the phenomenon-focused conception of *paradigm* in *Structure* suggest their own shift in the way we ought to think about the legitimacy of scientific problems and how that legitimacy is established and maintained. For the practitioner, the legitimate problems are a subset of the ones that can be worked on—that is, that can be put into a structure that makes them solvable. Those structures, schemata, frames—whatever you like—place certain phenomena in a specific kind of relation to certain other phenomena. Those phenomena receive their research legitimacy, I argue, not directly from how revealing they are about the secrets of nature, but rather from the fact that they are able to feature in a solvable structure. In rough form, the thesis I will be defending is this: a *legitimate* problem is one that can be set up in a way that makes it solvable. A problem gains legitimacy by wheedling its way into a solvable class—that is, through practitioners' efforts to categorize it as a member of certain kind of problem.

The frames, schemata, or structures that bestow legitimacy upon problems take a variety of forms, and they come in varying degrees of strength. I try to convey some feel for this variety in what follows. I do not, however, want to lose sight of the underlying functional unity that—oh, let's just call them *frames* from now on—that these frames possess: each of them in its own way facilitates the characterization of phenomena so as to incorporate them into a structure that makes problems solvable (Camp 2019b). It is this process of facilitation that, for example, impressed Kuhn in his discussion of how Galileo's model of the pendulum came to guide several subsequent, rather different types of investigations (see discussion in Chapter 2). Kuhn understood this process of facilitation in terms of "an acquired ability to see resemblances between apparently disparate problems" (Kuhn 1974, 471). There is no doubt that the process is sometimes driven by an acquired ability to see. But it is also, probably more often than not, driven by a concerted effort to treat some phenomenon as analogous to another—not because of perceived resemblances, but because *if we do so*, we will be able to characterize the latter in a way that produces a solvable problem. I provide a classic illustration of this gambit from the history of electricity in the next section.

That case is a fairly straightforward instance of the phenomenon I'm attempting to capture in this chapter. Things aren't always so simple. Following my discussion of electricity, I look in section 5.3 at an episode from the history of evolutionary paleontology in which the route from exemplar to frame was murky, circuitous, and protracted. Yet, in the end, a familiar outcome: a generalized means of structuring investigations so as to produce a result upon which other members of the field can build. The variety of responses to the exemplar in this case provides us with an important guide to understanding the types of responses that develop into a class of solvable problems.

In section 5.4, I step back from the details of particular research frames in the history of inquiry to consider some of the general properties of frames, properties that causally explain the production of multiple related classes—*families*—of solvable problems. This causal process, rooted fundamentally in practitioners' need to structure problems in a way that makes them solvable, captures what lies at the core of our concept of fruitfulness.

5.2. Maxwell's Mission

Observing the field of electrical research in 1855, Maxwell lamented a very particular kind of morass: "the present state of electrical science seems peculiarly unfavorable to speculation." At the heart of Maxwell's concerns was the need for a "simplification and reduction . . . of the results of previous investigation to a form in which the mind can grasp them" (Maxwell 1890, 155)—that is, some method that could "be of . . . use to experimental philosophers, in arranging and interpreting their results" (Maxwell 1890, 159). What he found particularly wanting was the lack of a method of investigation that allows the mind at every step to lay hold of a clear physical conception, without being committed to any theory founded on the physical science from which that conception is borrowed. (Maxwell 1890, 156)

As he explained, this situation was the result of two contributing factors. First, although a number of mathematical laws governing the behavior of electricity had been described, there was no unified mathematical framework from which those laws could be derived. Thus, "In order therefore to appreciate the requirements of the science, the student must make himself familiar with a considerable body of most intricate mathematics, the mere retention of which in the memory materially interferes with further progress" (Maxwell 1890, 155). One potential path out of the quagmire, he suggests, would be to try to capture all the known quantitative behavior using a "purely mathematical formula." The danger here, he felt, was that researchers might then be distracted by purely mathematical considerations, and thus "drawn aside from the subject in pursuit of analytical subtleties" (Maxwell 1890, 156):

> In [this] case we entirely lose sight of the phenomena to be explained; and though we may trace out the consequences of given laws, we can never obtain more extended views of the connexions of the subject. (Maxwell 1890, 155)

That is, while a purely analytical study could be counted on to reveal the far-reaching quantitative implications of a unifying mathematical framework, there was no guarantee that such a study would produce results whose potential physical meaning was clear enough to guide experiment.

On the other hand, there was no available substantive physical hypothesis as to the nature of electricity that could be safely used to guide further research without committing practitioners to experimentally unwarranted suppositions about what kind of thing electricity was. In the absence of a physical theory that was substantive enough to suggest how to proceed without exposing researchers to "that blindness to facts and rashness in assumption that a partial explanation encourages," practitioners were limited in their ability to formulate physical research problems that could be used to push forward the understanding of electricity (Maxwell 1890, 155–156). These two conditions—the absence of a unifying mathematical treatment of the known laws, and the absence of suggestive physical theory—together made for an intellectual landscape wholly inhospitable to inquiry.

We can obtain a clearer sense of Maxwell's concerns about the state of electrical science in his time by comparing them with that of the prior century, in which two of Kuhn's canonical exemplars, the *Principia* and the *Opticks*, had come to exert a variety of prescriptive influences on the field. Impressed by the success of Newton's inverse-square law, electricians of the eighteenth century sought to draw an analogy between *gravitational* attraction and *electrical attraction/repulsion*. Not a *physical* analogy, per se; it was generally assumed that electricity and gravity were physically distinct phenomena. Rather, they hoped to draw an analogy between the *form* of the force law for gravitational attraction and that of electrical attraction. This exploration would eventuate in Coulomb's Law, which is an inverse-square law. Crucial for our purposes here is the observation that Coulomb's Law is not a deductive consequence of Newton's law of gravitation, nor is the inverse-square form of the force law for electromagnetic attraction a deductive consequence of the form of the law of gravitation. What led electricians to explore the inverse-square form was simply the fact that it had worked for another kind of attractive force.

The history of these investigations reveals a set of constraints on inquiry that is significantly at odds with a certain standard picture of how scientific research works. The first recorded attempt to see whether electrical force might diminish with distance was made in 1746 by Christian Gottlieb Kratzenstein, who explored the possibility of the maximally simple inverse relation (1/r) (see Fig. 5.1). Here is Heilbron's account:

> He found that within what he called "experimentalerror"— 245 percent in one case—the force diminished inversely as the distance. . . . The approach was considered promising by such connoisseurs as Lambert and even Euler, who helped arrange Kratzenstein's appointment as "mechanicus" to the Petersburg Academy in 1748. Five years later, at the expiration of his contract, Kratzenstein happily accepted the chair of experimental physics at the University of Copenhagen. (Heilbron 1979, 460)

Could Kratzenstein's data exhibit the relation that Kuhn called *reasonable agreement* (Kuhn 1961)? Possibly—but reasonable *for what*? Clearly the agreement was not good enough for researchers to conclude that the problem had been solved (hence all the subsequent

Diſtantiae.	Adtraƈtiones	Calculus ex rat.ſimpl. inverſ. diſtant.
'''		
0	138 gr.	
18	38 *	38
38	19	18
45	13	$15\frac{5}{7}$
52	10	$13\frac{1}{8}$
56	7	$12\frac{1}{4}$
60	6	$11\frac{1}{7}$
71	5	$9\frac{1}{2}$
96	4.	$7\frac{1}{5}$

Figure 5.1 Kratzenstein's table of predicted and observed values for 1/r.

investigations). But it was ostensibly reasonable enough to warrant adopting the inverse relation as the basic frame for characterizing the law of electric attraction. Why was it that, despite the near-comical disagreement with observation, these investigations nevertheless convinced experts that it possessed enough promise to be adopted as a basic frame for structuring the study of electric force, and that they warranted a series of appointments to the most prestigious scientific institutions in Europe for Kratzenstein ("a clever man, able to discern interesting problems but unable to solve them") (Heilbron 1979, 485)?

What practitioners saw in Kratzenstein's experiment was not a correct expression of the electric force law, but rather evidence for the promise of a certain *analogy* with gravitational force as a way of structuring the problems of electric force. The table gives the unmistakable impression that *some* version of the inverse distance relation is likely to be quantitatively accurate. Recognizing that the experimental apparatus was still in its infancy, and that a simple inverse relation was not the only kind of inverse relation, Kratzenstein's experiment would have given them all they needed to be convinced that the analogy with gravitational force was a promising frame for structuring the study of electric force. Indeed, by 1760, Daniel Bernoulli had obtained results that could be modeled as an inverse-square relation.[3]

The other line of inquiry upon which eighteenth-century electricians drew was, of course, Newton's *Opticks*:

> Newton indicated that "a medium" might exist between . . . corpuscles; if this medium were an elastic aether, the sanction of the Master was available for the supposition of a kind of "elastic fluid" existing inside solid bodies, lodged as it were

---

[3] Roller and Roller (1950, 610). Heilbron (1979, 461) describes Bernoulli's result as "misleadingly similar to Coulomb's," because Bernoulli's set-up measures the tensile force of electricity, not "the electric force between elements of electrical matter" themselves (as does, say, Coloumb's torsion balance).

in the pores. Since Newton held the view that light (a "body" or "substance") entered bodies and was trapped within them, a Newtonian scientist—in the tradition of the *Opticks*—could well assume that the electric fluid or the caloric fluid of heat might also enter into bodies under proper conditions. Furthermore, the property of catching or trapping light varies, according to Newton, from one body to another because of structure and composition—the size and distribution of the corpuscles. So too, depending on the kind of body, there would be a more or less ready absorption of the electric fluid or caloric. Finally, the density of bodies was generally shown to be related to their optimal properties and we all see the attempt to relate electrical and thermal properties of bodies to their density. An exception was noted by Newton in bodies containing sulphureous or unctuous matter, and this too was relevant to theories of electricity and heat. Thus we may see at once how Newton's *Opticks* became a great stimulus to the growth of the speculative experimental physical sciences of heat and electricity. (Cohen 1956, 163)

This era of electrical science was favorable to "the growth of speculative experimental physical sciences of heat and electricity" in the sense that electricians had good reasons—"the sanction of the Master"—to structure their research in specific ways. With the imprimatur provided by Newton's expressed view that "light... entered bodies and was trapped within them," for example, electricians could with warrant investigate phenomena on the supposition that "electric fluid ... might also enter bodies under proper conditions." Just as Newton had argued that the structure and composition of a body affects the absorption of light, electricians saw themselves as justified in structuring their electrical research guided by the notion that such properties might also affect the absorption of electrical fluid. And so on. In this way, the *Opticks* served not only as an exemplar for the study of light, but for the study of electricity (and heat) through the exploitation of certain analogies that followed from representing the latter as a fluid. The *Opticks* thus fostered "speculation" in electrical science in the sense that it prescribed a framework for characterizing electrical phenomena and generating hypotheses that were susceptible to further empirical investigation.

Part of what made those electricians' use of the *Opticks* so productive was the very specific way in which they used it. While these men had confidently borrowed their conception of the behavior of fluids from Newton's penetrating insights into the behavior of light, their fluid-like conception of electricity did not itself commit them to any particular theory espoused in the *Opticks*. For them, the book could serve as a cutting-edge inventory of the essential or defining properties of subtle fluids, none of which was seen as dependent on Newton's optical theory per se. As such, taking a cue from Newton's investigation of how light might be carried through an effluvial medium did not necessarily result in framing hypotheses about electricity that would have been tainted with content specifically relevant to the study of light. The *Opticks* could be treated independently as either a masterful study of light, or, at a more abstract level, a clearly articulated exposition of the behavior of subtle fluids. By deriving guidance from this more abstract and less substantive physical framework, they were thus able to frame their investigations without adopting a specific "physical hypothesis" that might unduly confine them to "see the phenomena only through a medium, and [be] liable to that blindness to facts and rashness in assumption which a partial explanation encourages" (Maxwell 1890, 155–156).

Maxwell was in search of something that could function for our mathematical understanding of electricity as the inverse-square and fluid frames had functioned for the experimental investigation of electricity's physical behavior. This search could only conclude once he was in possession of "some method of investigation"— some frame—which could "bring before the mind, in a convenient and manageable form, those mathematical ideas which are necessary to the study of the phenomena of electricity."

His solution to this quandary was what he called a "physical analogy"—specifically, an analogy with fluids. By this, he meant a way of characterizing electricity that was *modeled* on the approach to representing fluids, "without being committed to any theory founded on the physical science from which that conception is borrowed," such as, for example, committing to the notion that electricity *was* a fluid. This approach to characterizing electricity was inspired by what he saw as the field's most fundamental need—viz., a way of representing simultaneously the direction and magnitude of electrical force at a given point. By Maxwell's time, electricians had learned enough about the quantitative behavior of electricity for him to see that a fluids-based approach would now be "capable of great simplification in the case in which the forces are such as can be explained by the hypothesis of attractions varying inversely as the square of the distance, such as those observed in electrical and magnetic phenomena" (Maxwell 1890, 159).

Maxwell's fluid-frame focused on a few salient properties of fluids which had been chosen specifically because of the ease with which they lent themselves to abstraction beyond a particular physical substrate. The opening salvo in his construction of the fluid-frame is bracing in its forthrightness:

> The substance here treated of must not be assumed to possess any of the properties of ordinary fluids except those of freedom of motion and resistance to compression. It is *not even a hypothetical fluid which is introduced to explain actual phenomena.* (Maxwell 1890, 160; my emphasis)

One can see in the emphasized portion how Maxwell is quick to forestall the mind's default tendency to begin visualizing a causal network in which a physically substantive electrical fluid is flowing around, producing electrical phenomena; as properties of fluids go, "freedom of motion" and "resistance to compression" are fairly innocent commitments from a physical point of view. Even so, he continues to clarify that the purpose of invoking these properties is emphatically *not* to ascribe physical qualities to electricity:

> It is merely a collection of imaginary properties which may be employed for establishing certain theorems in pure mathematics in a way more intelligible to many minds and more applicable to physical problems than that in which algebraic symbols alone are used. (Maxwell 1890, 160)

In other words, the properties "freedom of motion" and "resistance to compression" are intended only to be used as vehicles for bringing the well-established mathematical theory of fluids into contact with aspects of electricity that we want to represent—specifically, the direction and magnitude of electrical force (respectively). By 1855, the equations for motion and volume compression of fluids had a physical meaning for practitioners that was clear enough to see what they might suggest in the context of electricity, were it to be viewed from that perspective. Thus, for example, they could mathematically represent "freedom of motion" of electricity without ascribing to it any other qualities (specifically, substantive physical properties). As Maxwell expressed, the hope was that by showing practitioners how to frame aspects of electricity's behavior in terms of mathematics of fluid motion and incompressibility, "the results of previous investigations" would be rendered into "a form in which the mind can grasp them." Part of that "grasp" involved understanding what the mathematically derived behavior of fluids might look like in electrical contexts. The other part involved seeing the experimentally pursuable implications of that specific mathematical framing.

Key to Maxwell's strategy was the development of a characterization of fluids that was at once (a) *general enough* to be "applicable to any conceivable system of forces," (b) *specific enough* to facilitate physical theorizing, and (c) *precise enough* to imply quantitative results that could be investigated by experiment. The characterization's generality was achieved by constructing a conception of fluid that applied to anything capable of exhibiting directional motion and incompressibility *in some form*, regardless of the lower-level substantive physical characteristics.

5.3. Species-As-Particles

Maxwell's perspective on the fluid-framing of electricity as "not even hypothetical" represents one extreme position on a continuum of ways in which practitioners relate to the frames that they use to stabilize and advance inquiry. Much more recently, we can see a fascinating range of continuum positions occupied by members of the Marine Biological Laboratory (MBL) group in paleobiology during the 1970s. The core members of this group were David Raup, Thomas Schopf, and Stephen Jay Gould. Keen to apply the new technique of computer modeling to paleontological data, the group met at the MBL in Wood's Hole, Massachusetts, in December 1971. Initially their efforts went absolutely nowhere. Essentially, no one in the group seemed to have any sense of what would be a good question to ask the data that could then be answered by a computer (Sepkoski 2012, 222–223). On the verge of giving up, "and in some desperation in the final afternoon of the meeting," Raup suggested an exploratory experiment of sorts: to see what kinds of patterns of biological diversity would arise if all lineages had at a given time an equal probability of going extinct, speciating, and proceeding unaltered to the next moment in evolutionary history.

The answer they received was a big surprise: if lineages were randomly going extinct, diversity patterns would look pretty much the same as they do in the actual fossil record. Part of what the group members took away from this experience was that the shape of actual diversity patterns was not necessarily illustrative of the influence of natural selection, which, contrary to what the group's simulations assumed, nonrandomly favors some lineages over others on account of their relative fitness. Since those patterns could apparently also be generated through a random process, the patterns themselves were not sufficient to distinguish the influence of natural selection from that of random birth and death.

This seems to have been the extent of the substantive agreement within the group, for each member would go on to take very different perspectives on the significance of this approach for guiding future research. Schopf took the match between randomly simulated and actual fossil diversity patterns to be indicative of the fact that lineages *actually were* randomly going extinct, that natural selection was not responsible for the disappearance or the ascent of any lineage in evolutionary history. He came to view the history of life has more or less accidental, and his commitment to the literal truth of "species as particles" would only harden over time. As a result of this hardening perspective, he became increasingly invested in the idea that the right frame for characterizing research problems in macroevolution was something akin to statistical mechanics.

Raup took the match between the simulations and the fossil record itself to be of no great significance (the simulations were later shown to have made biologically unrealistic assumptions, which accounted for their closely approximating real fossil patterns). He certainly did not come to believe, as Schopf did, that extinction was a random process (see Raup 1991). But neither was he unaffected by the experience. It seemed to leave him with the impression that no one really understood the nature of extinction, and that theorizing about evolution on long timescales was not yet a well-governed endeavor. By all accounts, Raup was a master at suspending judgment concerning how the evolutionary process actually worked. He would gradually continue to update his perspective on the significance of random processes in shaping the history of life, driven largely by a talent for refining and clarifying the role of stochastic modeling as a tool for structuring research problems in paleontology. His interest in stochastic modeling derived not so much from a particular perspective on *nature*, but on *scientific inquiry*.

This perspective disposed him toward characterizations of research problems that tended to error on the side of well-substantiated conservatism. The more modest the frame, the more illuminating the results.

The effect of the MBL experience on Gould really defies description. On the one hand, he seems to have been attracted to the prospect of a significant amount of random noise in the evolutionary process.[4] He clearly believed that the ability to approximate actual diversity patterns through random simulations gave paleontology something distinctive to say about the evolutionary process, something which could only be seen by studying fossils.[5] On the other hand, he strongly resisted the main metaphysical lessons that Schopf had drawn (Sepkoski 2012, 263–269). Thus, the perspective he would go on to develop was unlike that of either Raup or Schopf. His interests lay not so much in advancing understanding in a responsible manner as they did in advancing the disciplinary stature of paleontology itself. He found the "stochasticity" frame compelling largely because it hypothetically excluded the influence of natural selection, thus differentiating the large-scale evolutionary processes that paleobiologists studied from the selection-saturated dynamics of microevolution (Haufe 2022). This is evinced by the fact that Gould largely abandoned the characterization of new research problems *tout court* after having briefly used the stochastic approach to help carve out a distinctively paleobiological approach to evolutionary theorizing.

The distinct ways in which each member of the MBL group related to the "species-as-particles" frame can be viewed as different positions on a continuum that tracks increasing degrees of literality. The details of the meeting in which it emerged make clear that the members' initial relation to that approach was as a *pretense*: the idea of simulating the history of lineages as a random walk was

---

[4] See, for example, Gould (1989) and also Beatty (1995).
[5] See Haufe (2015, 2022).

for no purpose other than to see what evolutionary history would look like under certain conditions which each member regarded as completely imaginary. No one at the time reflected any inkling that biological features of lineages literally made no difference to evolution. Their concerns were primarily methodological and exploratory, focused on how one might possibly introduce computer modeling techniques into paleontology. The random simulations were just one of a number of ways in which the group experimented with computing technology.

Unlike those other experiments, however, this one had a stabilizing effect on their future research which was almost immediate, and which would last throughout the remainder of each of their careers. Most interestingly, as noted earlier, this acute stabilizing effect took different forms for different members. For our purposes, Gould's is probably the least informative and the least interesting. Because of his overwhelming focus on saying something about evolution that was radical and distinctive, he would express different attitudes at different times. In correspondence with other MBL members, he maintains a healthy metaphysical distance from the "species-as-particles" frame. There is no indication that he took it literally. However, in print, he would often gesture at the radical implications of the random simulations for our understanding of the influence of natural selection, suggesting that the "species-as-particles" frame should be taken at least somewhat literally (e.g., Gould 1980). My assessment (elaborated in detail in Haufe 2022) is that Gould's primary interest lay in carving out a distinctive role for paleobiology as a contributor to the general theory of evolution, and he saw the promotion of the "species-as-particles" frame as the most promising way of doing that.

Schopf's example is precisely the reverse of Gould. The stabilizing effect on his work was just as immediate, but for metaphysical reasons with which he became increasingly preoccupied. The reason that "species-as-particles" had a stabilizing effect on him was because he took it as revelatory of the genuinely random nature of extinction. He seemed to find satisfying the prospect that there were no "inferior beings who deserved to die." Writing to Gould a few years after the MBL collaboration dissolved, Schopf spoke plainly: "it is now completely unacceptable to me that any *species* is really any different from any other species."[6] Far from facilitating disciplinary or professional ambitions, his literal orientation toward "species-as-particles" resulted in his increasing intellectual, social, and professional isolation, as Gould described in an obituary penned shortly thereafter:

> Tom was a prickly, often difficult colleague, so driven by his unconventional vision, so committed to its fundamental truth, so brave (or foolhardy) that he would sacrifice friendship and human relations to its zealous advance. . . . I know that it led him to much personal misery. . . . Tom was so committed to a unity of vision that he hopelessly conflated his sense of the factual with his belief about the ethical. He saw what he deeply believed as not only true but just, right, and moral.[7]

Clearly, "species-as-particles" was no pretense for Tom Schopf.

The frame's stabilizing effect on Dave Raup's research best captures the fundamental epistemic phenomenon with which I have been most concerned in this book. For Raup, "species-as-particles" persisted as a pretense for the rest of his career, albeit of a very specific sort. His cautious development of the frame throughout the 1970s and 1980s led him eventually to adopt a suitably refined version of it as the appropriate framework for the characterization of evolutionary systems in the absence of directional change—that is, the appropriate "null model" against which to compare causal hypotheses in paleobiology (see also Gould 1978). By creating an imaginary world in which no evolutionary forces were acting, Raup

---

[6] Quoted in Sepkoski (2012, 266).

[7] Gould (1984, 282). Quoted in Sepkoski (2012, 268–269).

thought, we could distinguish between what kinds of fossil patterns were indicative of the influence of evolutionary *forces*, and what was just nondirectional meandering of lineages through morphospace. To see whether evolutionary forces like natural selection were operating on the historical development of a lineage, one would begin by pretending that *no* such forces were acting, and then compare the resultant pattern to the actual fossil record. If there was little or no deviation from the null expectation, the investigator lacked a basis for inferring the influence of evolutionary forces. If the deviation from the null expectation was significant enough, the investigator would then be justified in pursuing causal explanations for the actual fossil pattern.

Now, the relatively immediate effect of the "species-as-particles" approach on paleobiology presents something of a puzzle. The MBL papers, through which the approach was first promulgated, reported no empirical discoveries. They reported no theoretical results of any great significance. Most importantly, the specific form in which the frame was instantiated was quickly proven to be fundamentally misguided (Stanely 1979; Stanely et al. 1981; see Chapter 8 for discussion). If the MBL papers were known within a very short time of their publication to tell us nothing about nature, why was their "species-as-particles" frame so influential?

The answer lies in how "species-as-particles" facilitated the characterization of randomness in evolutionary history. For reasons that are readily explicable, the particular effect that the "species-as-particles" frame had on Raup turned out to be the one which would fundamentally transform paleobiological research. Paleobiologists hoped to make causal claims about the evolutionary history of lineages, and they immediately saw "species-as-particles" as the vehicle through which they could generate the characterizations of randomness they needed for formulating and testing causal models. It did not matter that the *particular* characterization of randomness represented in the MBL papers was biologically unrealistic. Other researchers saw that, with a few simple tweaks, they could produce a mathematically and biologically plausible characterization of a zero-force state for macroevolution, a characterization which they could use as a null/neutral model. The MBL papers themselves failed to provide an apt characterization of randomness. But they succeeded in giving birth to the frame within which an acceptable characterization of macroevolutionary randomness could be realized. This frame continues to the present day to function as the default means by which paleobiologists endeavor to structure research problems (Sepkoski and Ruse 2009).

The "species-as-particles" episode nicely illustrates Elisabeth Camp's observation that "different scientists will often bring markedly different characterizations and perspectives on their subjects to the interpretive table, especially at the beginning of inquiry." But it also points to a more epistemologically significant theme—viz.,

that frames do more than just interpret a fixed set of assumptions about their targets: they provide open-ended tools for assimilating new information and for generating hypotheses about undiscovered features and causal structures. (Camp 2019a, 330)

In this case, the frame led to research questions which have profoundly reshaped our understanding of the evolutionary process. Sensitive to the nearly spontaneous and geographically widespread appearance of vastly different forms that often follows an equally spontaneous disappearance of many forms across the globe, paleontologists for generations had suspected that mass extinction events are an important driver of evolutionary diversity. Inspired by the approach to framing paleobiological research questions exemplified in the MBL papers, a young researcher named Jack Sepkoski asked, "What would the history of Phanerozoic diversity look like if lineages were randomly branching and going extinct, with the occasional large-scale decimation of lineages corresponding to the actual mass extinction events?" It turns out that it would look pretty much the way it actually does (Sepkoski 1978,

1979, 1984). After a mass extinction, diversity levels seem to "recover" to or beyond their pre-extinction levels due merely to the availability of ecological space. This realization led to closer examination of specific recovery periods, which disclosed really significant differences in the way in which the specific scope of an extinction event can affect the degree of evolutionary novelty that emerges during the recovery period; unless the extinction event clears out "adaptive zones," fundamental evolutionary novelty tends not to emerge (Erwin et al. 1987).

This is just a small sample of the myriad lines of inquiry that have been generated through the "species-as-particles" frame. And it is through the ability to pose and investigate questions like these that paleobiology has become our most valuable tool for understanding evolutionary diversity and innovation. The episode also illustrates a related epistemologically significant theme, one that will occupy center stage in the next chapter: the *unforeseeable effects* of frame adoption. In adopting the "species-as-particles" approach as the appropriate frame for zero-force evolutionary states, paleobiologists inadvertently transformed their understanding of randomness itself. Random evolutionary walks do not lead to nice uniform distributions of traits as one might expect; they tend to skew significantly. In paleobiologist David Jablonski's words, "Randomness is clumpy."[8]

Knowledge of the clumpiness of randomness would, in turn, dramatically alter practitioners' perspective on evolutionary disparity. It is obvious to anyone who looks at the natural world that the distribution of forms is itself clumpy—quite clumpy. Real organisms do not exhibit all physically possible ways in which organisms could be. There are large groups whose members resemble each other to varying degrees, and nothing in the morphospace between groups. In a more innocent time, it would have been natural to see these clumps as evidence of the optimizing power of natural selection.

[8] Personal communication.

But the realization that randomness itself is clumpy caused a shift in perspective on how to interpret the clumpy and sparsely occupied morphospace. The disposition to characterize clumps as selective optima gave way to a disposition to characterize them as potential absorbing boundaries in a random walk.

The shift in perspective on randomness nicely illustrates core features of what Elisabeth Camp (2019a) calls a *characterization*. A characterization is an application of a frame to a particular subject. To "apply" a frame to a subject is to "structure our intuitive thinking about a subject," which involves:

- Assigning degrees of significance to certain features of experience—that is, making them *salient* to different degrees
- Assigning relationships between salient features (Camp 2019a: 308–309).

It is easy to see this dynamic reflected in the MBL revolution in paleobiology. The shape of a clade's diversity over time typically takes the form of a spindle, and cladograms are often referred to as "spindle diagrams." This spindly shape had traditionally been treated as salient, because of its presumed causal connection to natural selection. The thinking was that a clade is "born" in a state of low diversity (one lineage). As it becomes better adapted to its environment, it begins to flourish, expanding its number of lineages. At some point, due to abiotic environmental change or to habitat invasion, its numbers begin to decline. Eventually, the clade's remaining lineage loses its battle with environmental challenges, and the clade becomes extinct.

It seems that, from the moment the MBL group saw that these spindles could be generated through random simulations, the spindle shape of a clade's diversity lost all significance (Huss 2009). This is precisely the moment at which we see the group members begin to undergo the transition that would eventuate in the discipline-wide adoption of the "species-as-particles" frame. For, if spindles could be produced through pure chance, then the spindle-shaped diversity was not immediately attributable to natural selection. Apparently, clades could flourish through chance. They could decline through chance. They—even highly diverse clades—could go extinct through chance. The immediate effect of this realization for Raup would be to alert him to the distinct possibility that evolutionary theory was devoid of any true understanding of extinction: "I am becoming more and more convinced that the key gap in our thinking for the last 125 years is the nature of extinction" (Sepkoski 2012, 259). In its mature, Raupian form, the shift to the "species-as-particles" frame marked the emergence of a *science* of extinction. But spindliness could no longer be seen as evolutionarily significant.

The salience of clumpiness in morphospace went through a distinct but related shift. On the one hand, the MBL random simulation of clumpiness led to a similar shift in perspective on the *meaning* of clumpiness. Like spindles, clumps in morphospace could no longer be treated as solid evidence for selection. But clumps did not thereby cease to be salient. In fact, if anything, clumps in morphospace are *more* salient now than they ever have been. The effect of the ability to randomly simulate clumpiness was to elevate its mystique, issuing challenges to researchers to develop a new understanding of how clumps actually emerge in evolutionary history. In essence, this is similar to the transition that spindliness underwent. Spindliness is an artifact of a particular set of choices about how to visually represent data on the origination and extinction of lineages. As with clumpiness, the realization that spindliness could be randomly generated resulted in a similar elevation in the mystique of origination and extinction of clades. In this way, both clumpiness and origination/extinction went from solved (or non-) problems to mysteries in need of solution.

Consistent with Camp's account, the "species-as-particles" frame led to a restructuring of research problems in paleobiology by (a) *re*assigning the relationships between certain features (neither clumpiness nor spindliness is any longer directly connected to natural selection), and (2) *re*assigning the degree of salience for certain features (spindliness is no longer salient; clumpiness is way *more* salient)—what she might call a *re*characterization.

We can also see in the MBL episode how the perception of salience led to the emergence of an exemplar. The first set of randomly simulated clade diagrams produced at Woods Hole instantly grabbed the group members' attention. In line with last chapter's discussion, what was salient about these diagrams was their resemblance to cladograms that were taken to be emblematic of natural selection's influence on a clade's history. The salience of this resemblance only existed against a backdrop of training and a disciplinary tradition of a specific form of representing taxonomic diversity. Spindliness is not intrinsically significant. It was salient to paleontologists because of their training and experience. The immediate result of the perception of salience in this case was the minting of a new kind of problem structure in evolutionary biology, characterized by the use of random simulations to serve as null models against which to frame causal hypotheses about evolutionary history. Indeed, the second MBL paper, published within a year of the first, applied random simulations to the evolutionary history of morphological structures, an important step in the development of investigations of clumpiness in morphospace (Raup and Gould 1974). We can thus see how, as argued in the previous chapter, the initial perception of salience led to a disposition to characterize other subjects as members of the same class of problems, a class characterized by a novel perspective on (1) what evolutionary randomness looks like and on (2) how to understand actual fossil evidence in light of what evolutionary randomness looks like. It is in virtue of serving as the basis for this disposition that the set of randomly generated clades from the first Woods Hole meeting constituted an exemplar. Their status as exemplar would only solidify further as the rest of the discipline's practitioners found themselves similarly affected by the "species-as-particles" frame, to which they were introduced by the MBL's publications during the 1970s.

The MBL episode provides a clear example of a real-life instance of exemplar emergence, one which closely follows the abstract process described in the previous chapter. That process began with an initial detection of salience produced by the perception of similarity between real and random clades. It also provides a clear illustration of how Camp's (2019a) taxonomy elegantly parses the major features of research experience. The events at Woods Hole led to a shift in *perspective*—in many ways a gestalt shift; where the group's members formerly saw the imprint of natural selection, they now saw the imprint of chance (or its potential). This new perspective was "crystalized" in the "species-as-particles" frame, which allowed researchers to approach evolutionary history *as if* lineages were randomly branching and going extinct. The frame facilitated this approach by serving as a guide for *characterizing* zero-force evolutionary states. While the MBL group members agreed on this much, in practice they related to the frame in different ways, ranging from the literal (Schopf), to the pretend (Raup), to something amorphous (Gould). As the years went by, paleobiologists became increasingly interested in structuring their investigations on the model of the MBL approach.

5.4. The Normative Significance of Frames

As usual, I find it useful here to set the direction and tone of a new discussion by outlining the spirit of Kuhn's account in *Structure*, emphasize where it is misguided or insufficient, and then press on with my own modest yet critical amendments to the picture of scientific decision making that he presents. Recall that this account is anchored in Kuhn's commitment to the notion that research exemplars—"achievements sufficiently unprecedented to attract an enduring group of adherents away from competing modes of scientific activity"—are solved problems. Hopefully by now the idea that exemplars need to come in the form of solved problems has been thoroughly debunked. But, like so many other debunkings, this one comes at a cost in the form of whatever understanding we thought we'd achieved on the basis of that debunked idea. In this case, our impassioned debunking efforts have resulted in the loss of an understanding of the source from which a paradigm derives its normative powers—its powers to strike fear into the hearts of researchers were they even to *think* about stepping beyond its narrowly prescribed boundaries.

While I am satisfied that no praise would be wasted on Kuhn's depiction of the many dimensions of inquiry along which a paradigm exercises normative influence, it cannot be said that Kuhn invests an excessive amount of time into explaining just how these normative powers flow from exemplar to paradigm to researcher. However, some basic features of his view can be pieced together from two sources: (1) his insistence that exemplars are solved problems; and (2) his view that, in the minds of practitioners, the ensuing paradigm provides a more or less correct picture of some aspect of the world, reflected in such statements as this:

> Normal science, the activity in which most scientists inevitably
> spend almost all their time, is predicated on the assumption that
> the scientific community knows what the world is like. Much of
> the success of the enterprise derives from the community's will-
> ingness to defend that assumption, if necessary at considerable
> cost. (Kuhn 1962/2012, 5)

Together, (1) and (2) suggest a canonical source of normativity: truth. Normal science is predicated, he says, on the assumption that community knows what the world is like. From where did they get that knowledge? From the exemplars, the problem-solutions deemed to have solved a problem by revealing some truths about nature. For the practitioner, Kuhn's paradigms institutionalize these truths, and that institutionalization has effects on her thinking that can go as deep as you like, ranging from her beliefs about what is true or probable to the very theory-laden perceptions that compose her "world." She can no more deny the constituent posits of the paradigm than she can the deliveries of her own senses. The paradigmatic picture is *the* criterion against which both propositions and experiences are measured. That is normative significance itself, is it not?

The challenge we now face is to find a way of accommodating the deferential behavior of the practitioner at the throne of the paradigm without necessarily appealing to her beliefs about the natural world or the theory-laden content of her perceptual experiences.[9] I have argued that solved problems are but a single subvariety of research exemplar, perhaps even a marginal one. They are part of a genus of problem-solving attempts that exemplify an approach to setting up research problems so as to make them solvable. That genus contains other subvarieties, other kinds of problem-solving attempts that cannot claim uniform success, but which nevertheless were taken to exemplify an approach to framing solvable problems; subvarieties like the kind represented by Maxwell's fluid frame, which he explicitly advertised as a pretense; subvarieties that include acknowledged failures, like the biologically unrealistic MBL papers described earlier; subvarieties like Galileo's *Discourse on the Two New Sciences*—which is a mixture of successes (such as his proof that the distance traveled under uniform acceleration is proportional to the square of the time traveled) and failures. Indeed, much of the "mop-up" work left by an "achievement sufficiently unprecedented" is mopping up a mess—developing failed problem-solving efforts into solved problems. The normal scientific activity that grew out of Maxwell's "On Faraday's Lines of Force" cannot

---

[9] Readers will recall that Kuhn was "impressed with the implausability of the view" that scientific theories approach truth (Kuhn 1962/2012, 205). That view may well be false. Were it so, it still would not pose a problem for practitioners of the natural sciences believing in the truth of their theories (although Laudan 1977 thinks it does).

credibly be described as "predicated on the assumption that the scientific community knows what the world is like," nor could that which grew out of the MBL papers. Come to think of it, neither can premodern astronomy—one of Kuhn's "go-to" exemplars of mature science (Kuhn 1962/2012, 68–69). The community of Ptolemaic astronomers did not claim to "know what the world is like," aside from their belief in geocentrism. The modeling system that defined that community—that, by Kuhn's lights made it a mature scientific community—was in no way taken to be an expression of their knowledge of the physical world; to which part of "what the world is like" does the equant correspond? Taken together, these examples strongly indicate that the source of a paradigm's normativity is neither the presumed truth of certain propositions regarding the natural world nor the authority of sense perception. For, we now know that paradigms exercise their normative influence over research practice in the absence of both of these.

In place of the normativity of truth, let me offer an alternative: practitioners humble themselves before whatever they believe will allow them to structure research problems so as to make them solvable. Normal science, a phenomenon of supreme importance ably described by Kuhn (Hacking 2012), is predicated on whatever lifeline practitioners end up grasping as they desperately grope for a vehicle with which to make progress on questions that interest them. Paradigms/frames offer practitioners an answer to the all-important question, "What ought I to work on?" because they define and elucidate the kinds of problems that can be solved. Kuhn clearly understood that paradigms define the legitimate problems for a field. Where I believe he erred was in thinking that the source of a problem's legitimacy was its consistency with or relation to some picture of the world, rather than its solvability. Kuhn never tells us how what makes a paradigm's "mop-up work" possible. It would be a small favor indeed if exemplars provided nothing more to practitioners than a general picture of the world which highlighted a number of significant gaps in our knowledge, none of which we had the means to plug. On the other hand, if the only thing to which an exemplar could claim credit was that it was able to serve as a sample of problem solving by providing a schema for structuring investigation, it will have repaid in full whatever attention researchers had given it.

The idea that the normative significance of a frame lies in what it tells researchers about the kinds of problems that are solvable is more modest than the "worlds" that Kuhn offers them, but it is more explanatory and it is more consistent with evidence from the history of science. It is more explanatory in this sense that it by itself can explain the massive proliferation in successful research following a paradigm shift; by contrast, a change in worldview alone would be insufficient to achieve this. That it is more consistent with the evidence provided by the history of science, I believe, warrants no further elaboration. That is what most of this book has been about.

It has also been useful at times to gesture toward the fact that epistemological picture I am offering is also more general than the one Kuhn provides, in that it is able to connect the kinds of decision-making we see in the natural sciences with those in mathematics. Things like mathematical objects, notation, and diagrams, for example, all share the important property being uncontroversially subject to the discretion of the practitioner (Muntersbjorn 2003). There is no correct system of notation hiding out there for us to discover, just as there is no correct diagrammatic depiction of forces. There arguably are mathematical objects awaiting discovery, but none of these objects is ontologically privileged over others in the way we often think of, say, the collection of particles composing my desk as being ontologically privileged over the collection of particles composing my desk and the bus now rattling past my house. When a mathematician develops an object, or a definition, or a system of notation, there is no remaining question of whether she has things right; all that matters is whether and to what degree these artifacts facilitate inquiry. Phenomena like these, drawn from mathematics, offer us an opportunity to examine the manner in which practitioners make choices about how to structure inquiry in an investigative context in which questions about correct taxonomies cannot possibly arise.

Why is such an opportunity valuable? Plainly, because we want to see what drives practitioners' choices when scientific realists are not looking over their shoulders. We want to know what scientific practice would look like if practitioners were unburdened by the notion that their charge is to carve up experience as nature intended. Mathematics affords many such opportunities, because issues about correctness arise in relatively few investigative contexts. No one begrudges the mathematician his choice among acceptable proof styles, his preference for a geometrical approach over an arithmetical one, or his use of an object or notation of his own invention. Mathematicians perennially find themselves facing the choice among a number of proofs of the same theorem. These proofs do not differ in their degree of correctness. They do, however, differ in many other important respects, and mathematicians characteristically express a preference for this proof over that one, preferences which they defend on epistemic grounds: this proof offers a better understanding of the theorem; that proof displays the sense in which the theorem's truth was "inevitable"; still another proof highlights the relationship between this theorem and some other important result. While these considerations often have an aesthetic component to them, the motivation for employing them cannot be strictly aesthetic. Choices made on the basis of these considerations and others like them are what drive the growth of mathematical knowledge, and on that basis they are properly classified as *epistemic* considerations. Certainly mathematicians regard them as such. They do not characteristically take choices driven by these considerations as mere expressions of mathematical taste. Rather, they regard such considerations as important for *mathematics.*

Mathematical practice thus gives us a clear illustration of how certain choice criteria can fuel the growth of knowledge without servicing the goal of truth. This was a feature of deep interest to Henri Poincaré, whose reflections on mathematical practice emphasize it. His essay, "The Future of Mathematics," is a paean to the principle that the solution to a mathematical problem is of no value unless it can be treated as an exemplar.

> Suppose I have undertaken a complicated calculation and laboriously reached a result: I shall not be compensated for my trouble if thereby I have not become capable of foreseeing the results of other analogous calculations and guiding them with a certainty that avoids the gropings to which one must be resigned in a first attempt. On the other hand, I shall not have wasted my time if these gropings themselves have ended by revealing to me the profound analogy of the problem just treated with a much more extended class of other problems; if they have shown me at once the resemblances and differences of these, if in a word they have made me perceive the possibility of a generalization. (Poincaré 1910, 79)

Notice here that the value of the result lies not in whether it is correct, but in whether it has revealed "the profound analogy of the problem just treated with a much more extended class of other problems." Such a revelation does not depend upon the result's being correct. It rather depends upon whether the mathematician is able to project aspects of his approach to the focal problem onto other phenomena of interest—whether he is able to build a bridge between this problem and others which suggests that their solutions might be related. Again, with prejudice:

> A new result is of value, if at all, when in unifying elements long known but hitherto separate and seeming strangers one to another it suddenly introduces order where apparently disorder

reigned. It then permits us to see at a glance each of these elements and its place in the assemblage. This new fact is not merely precious by itself, but it alone gives value to all the old facts it combines. (Poincaré 1910, 80)

Poincaré goes out of his way in this essay to emphasize just how insignificant the correctness per se of a mathematical result is, on account of the fact that the correctness per se offers almost nothing to our efforts to pursue other research questions:

> when a rather long calculation has led to some simple and striking result, we are not satisfied until we have shown that we should have been able to foresee, if not this entire result, at least its most characteristic traits. Why? What prevents our being content with a calculation which has told us, it seems, all we wished to know? *It is because, in analogous cases, the long calculation might not again avail, and that this is not so about the reasoning often half intuitive which would have enabled us to foresee.* This reasoning being short, we see at a single glance all its parts, so that we immediately perceive what must be changed to adapt it to all the problems of the same nature which can occur. (Poincaré 1910, 81)

In the standard case, what we are seeking is not simply the solution to a specific mathematical problem, but the discovery of a *family* of problems, defined by certain "characteristic traits" which are exemplified in the one we've just attempted to solve, and which we can use to structure our approach to other problems. Whether a property of the result counts as a "characteristic" or not will depend on whether practitioners can be persuaded to treat aspects of other problems as versions of that property. As every taxonomist knows, new specimens do not come with instructions detailing what their characteristic traits are. A trait is elevated to the level of "characteristic" through the discovery that it best facilitates classifications that are dispositive of other significant, discipline-specific questions.

As a powerful illustration of this theme, Rudwick (1972) describes how early attempts to even *draw* fossils were hampered by a lack of guidance as to which features of the fossil were biologically significant and which were not (or were not even part of the organism).

Conclusion

An analysis of the rationality of inquiry will always favor truth as pre-theoretically the most plausible driver of decisions made in the pursuit of knowledge. Acknowledged truths force themselves into our corpus of beliefs in a unique and profound way. Nor can we deny the deep connection that the concepts of truth and knowledge bear to one another. Can a search for knowledge be at all credible if it is even slightly unconcerned with truth, or if it is willing to put other values above truth? Is there any plausible reading of the sentence, "I know that p and p is not true," or of the sentence, "I know that p but I do not believe that p," that is not obviously self-contradictory? Any attempt to deviate from this stable and well-behaved cluster of notions should be met with extreme skepticism.

But the practice of science and the goals in furtherance of which is practiced are a good deal more complicated than can be grasped by thinking of them as an instance of the sort of epistemic phenomenon to which these admittedly compelling considerations apply. The scientist's desire to know resembles that of the nonscientist in every meaningful respect: he wants to know what is true of nature. However closely these states are aligned, what satisfies that desire in the context of scientific inquiry is something that does not admit classification as readily or as tidily as that of the idealized epistemic agent who features routinely in philosophical analyses of knowledge. Nor does it relate to the canons of rationality in the ways we have been taught to expect through the examination of epistemic phenomena far removed from scientific inquiry. There are pressures that bear directly on the pursuit of scientific knowledge that have no plausible role in the pursuit of truth. These pressures are not on the margins. They are the beating heart of practice. To understand the shape of scientific rationality, we need a different epistemological toolkit.

Any toolkit adequate to the task will need to pick up the explanatory burden for which truth has traditionally been responsible. Only part of that burden consists of trying to understand what has made science so successful. The other part consists of trying to understanding what other than truth could possess a normative stature sufficient to influence scientific decision making. The many problem-solving capacities facilitated by the ordering systems that I have called frames give us a sense of how important they must be to the practicing scientist. Simply put, they allow him to practice science. In so doing, frames offer "the actualization of a promise" that truth cannot reliably fulfill. This is what grounds their claim to normative significance.

6

Fruitful Classification

NPR Host: Leslie, could one say that a book is nothing more than a painting of words, which are the notes on the tapestry of the greatest film ever sculpted?

Leslie Knope: One could say that. But *should* one?

Parks and Rec, "Born and Raised"

6.1. Promise Kept

When it came to research, Poincaré made no secret of the fact that he was primarily only interested in a really big score. He didn't just want to rob a bank. He didn't just want to rob a whole bunch of banks. He wanted to have his own mint. Poincaré's reflections are the most direct and explicit testimony we have so far encountered on the connection between promise and fruitfulness. I want to draw out two dominant themes from his discussion, each of which bears on a different dimension of the priority of promise in scientific decision-making.

Let's begin by relating Poincaré's comments on the value of an experiment or of a mathematical proof to our earlier discussion of promise. One thing he makes clear is that he is much less interested in finding specific answers than he is in finding a means with which to do more research—that is, something that can be applied in other contexts to shed light on hitherto unexamined problems. Individual bits of research acquire normative import when they are able to serve as such a means. Practitioners,

Fruitfulness. Chris Haufe, Oxford University Press. © Oxford University Press 2024.
DOI: 10.1093/oso/9780197666395.003.0006

perennially in need of something to work on, will have strong motivation to take those individual bits seriously; they *ought* to use them to further their own research if they are interested in solving problems, because they will then have a general-purpose toolkit for raising and solving problems. In this way, those individual bits "promise the concrete successes for which scientists are ordinarily rewarded." In Poincaré's musings we can, for example, see clear reflections both of Maxwell's lamentation and its amelioration in the form of the "fluid" frame for electricity. His perspective also explains why the MBL papers enjoyed the particular history that they did. Those papers contributed far more to our ability to *pose and investigate questions* about the evolutionary process than they did to our understanding of the evolutionary process itself.

The question now before us is how the acquisition of a framework for structuring research problems connects to the notion of fruitfulness per se. Up until now, our discussion has focused on the process by which a class of solvable problems emerges from an exemplar and on the value that practitioners naturally place on any such class. Something's missing, though. Because, at the heart of the notion of fruitfulness is the emergence of *multiple kinds of benefits*, many of which were unforeseeable at the outset. For example, when I say that a meeting was fruitful, I don't just mean that it was productive or beneficial. A meeting doesn't have to be fruitful to be productive. If we arrive at the meeting, accomplish what we set out to accomplish, and part—not necessarily amicably—then, we've had a productive meeting. A fruitful meeting requires something more. If I report that a meeting or discussion I had earlier was *fruitful*, I mean to indicate that some unexpected benefit came out of it, that things had gone in multiple directions that were not foreseeable, and that each of those directions was valuable in its own way. It also suggests that things were left in a somewhat pregnant state when our discussion ended, and that we'd opened up lots of possibilities for further development. Fruitful meetings and discussions don't need to resolve anything, or even begin the process of resolution. By contrast, I can have a productive discussion with one of my enemies; as long as we make some progress toward resolving an issue on which we oppose one another, that discussion was productive even if we are certain to remain enemies after the issue has been fully resolved. Likewise, a *fruitful* endeavor is not merely productive; it is not even necessarily productive. If I describe, say, this evening's walk as productive, normally all that I mean is that I got out of it what I wanted. You wouldn't say, "Really? How so?" However, if I describe the walk as *fruitful*, you might naturally follow up my comment with an inquiry into what made it especially fruitful, what happened that I hadn't anticipated, how it differed from my typical evening constitutional. The word *fruitful* is carefully chosen to mean something quite distinct. It's not a word I use lightly.

The picture I've developed up to now has yet to account for these essential ingredients. That picture has given us an understanding of how a class of solvable problems emerges, but it has shown us nothing with respect to the tantalizing variety, unforeseeability, or hereditary relatedness that is central to the notion of fruitfulness. And until it does, it cannot properly be considered an account of fruitfulness in science, nor can it fully resolve the Puzzle of Promise. A fruitful problem is, in Poincaré's words, one that "reveal[s] to me . . . a much more extended class of other problems"; Hilbert describes it as a "guide post on the mazy path to hidden truths" (Hilbert 1902, 438). Similarly, we find Frege insisting that a "fruitful definition" in mathematics must possess the element of surprise: "What we shall be able to infer from it, cannot be inspected in advance" (Frege 1980, 100–101). Fruitfulness in the context of research is not solely about the production of solvable problems. It is about the *unexpected* production of *multiple families* of solvable problems.

In this chapter I complete the promised account of fruitfulness by drawing a straight line from the emergence of an exemplar to the development of a propensity to generate families of solvable problems. These events constitute different stages in a single process that is fueled by the intellectual conservativeness of research communities—that is, their reluctance to abandon an incumbent research framework. Kuhn understood this; indeed, it is one of the most surprising and most powerful features of his model that he is able to provide a compelling account of how the drive to avoid novelty literally causes the emergence of novel scientific discoveries. In *Structure*, the fact that "Normal science does not aim at novelties of fact or theory" was fundamentally rooted in the research community's belief that they know more or less what the world is like. Thus, to adopt novelty of fact or theory as a goal would be to undermine the very possibility of normal science itself, which is, according to him, premised on practitioners' assumption that the incumbent theory is physically correct.

I believe that we now have ample justification for rejecting this view as a generally correct description of what motivates the research community to doggedly adhere to a certain way of solving problems. In place of this view, I advocated an alternative account of the normative significance of a frame that locates its motivating power in the simple fact that it provides problems with a certain structure that makes them amenable to solution. This simple fact is sufficient to explain why researchers continue to employ a frame once it has arisen. And the historically low frequency with which such opportunities arise is reason enough to explain their reluctance to abandon a frame. We do not lose the ability to explain the avoidance of novelty just because we dispense with Kuhn's overly strong account.

More importantly, as I show in sections 6.2 and 6.3, we gain a considerable amount of explanatory power that Kuhn's model never offered us with respect to the *proliferation* of novelties. It's no secret that Kuhn's reliance on the significance of practitioners' perceptual capacities for explaining paradigm shifts was not overly popular. In its place, I offer a much weaker set of premises that gives us the kind of novelty production that Kuhn thought of as beholden to changes in worldview, as well as a formidable array of more modest results. Moreover, it provides an explanation for why a Kuhnian change in worldview occurs simultaneously with a method for structuring research problems such as to make them solvable. There is no reason why this should occur in Kuhn's model. I show why it is inevitable.

The central feature of my alternative model is the cognitive behavior of metaphor. Consistent with other developments in his thinking after *Structure*, Kuhn appeals to the fact that practitioners make use of metaphors and analogies in their research, as in this lovely passage:

a good deal has been said in my original text under such rubrics as "metaphysical paradigms" or "the metaphysical parts of paradigms." I have in mind shared commitments to such beliefs as: heat is the kinetic energy of the constituent parts of bodies; all perceptible phenomena are due to the interaction of qualitatively neutral atoms in the void, or, alternatively, to matter and force, or to fields. Rewriting the book now I would describe such commitments as beliefs in particular models, and I would expand the category models to include also the relatively heuristic variety: the electric circuit may be regarded as a steady-state hydrodynamic system; the molecules of a gas behave like tiny elastic billiard balls in random motion. Though the strength of group commitment varies, with non-trivial consequences, along the spectrum from heuristic to ontological models, all models have similar functions. Among other things they supply the group with preferred or permissible analogies and metaphors. By doing so they help to determine what will be accepted as an explanation and as a puzzle-solution; conversely, they assist in the

determination of the roster of unsolved puzzles and in the evalu-
ation of the importance of each. (Kuhn 1962/2012, 183)

Had Kuhn rewritten *Structure* in the way here envisioned, he would
have found himself with few resources to support the normative
influence of paradigms; the ungainly phrase "beliefs in particular
models" is a reflection of the fact that he had not sufficiently thought
this through. Moreover, Kuhn says almost nothing about how
models "supply the group with preferred or permissible analogies
and metaphors"; what little he does say focuses on practitioners
"learning to see" in a sense that never moved beyond base perception,
resemblances, and learned similarity relations. In contrast, I focus
on a much broader range of supply lines, primarily those which
practitioners consciously and strategically impose on problems—or
what I'll call *stabilizing strategies*. The effects of stabilizing strategies
are explicable in terms of the well-known effects of metaphor, which
in turn explain many of the important features of paradigm shifts for
which Kuhn's "shifts in vision" provide an exceedingly poor account.

I'm going to begin by developing the second of the two themes
that emerge out of Poincaré's discussions on fruitfulness—viz.,
the use of analogy. Building on his ideas, I make some general and
astute observations about the role of analogy in the art of classification. With the basic features of analogy in place, I describe
a well-defined sequence in research practice in which research
communities move from perceptually driven acts of classification to the more fiat-fueled ventures that I've been emphasizing
throughout this book. It turns out that the latter have unwieldy
cognitive effects that are able to account for the emergent novelties
that interested Kuhn, and much more besides. Not only that, they
provide a single unifying explanation for many of the phenomena
associated with scientific decision-making, such as its uniformity
and simultaneity. I conclude with some preliminary remarks on the
next chapter's topic of what it means for the use of certain analogies
and metaphors to be justified.

6.2. Fruitful Analogies

Kuhn was wont to portray all classification as a reflex of the mind governed by theory-laden perceptions and centered on "acquired similarity relations." It is congenial to his general picture of paradigm choice and it fits perfectly with his effort to deprive paradigm choices of a rational basis. But there is a part of me that finds Kuhn's attraction to this view completely incomprehensible. It does not fit well with the overall heterodoxy of *Structure*, nor with the generally pragmatic orientation of his philosophy of science. In a beautiful essay on Kuhn's thinking leading up to *Structure*, Peter Galison (2016) reveals a man utterly enamored of the dominant psychological theories of the day, which were really just coming to grips with the theory-ladenness of perception. Kuhn clearly loved this work, investing studies like Gerald Postman's "anomalous playing cards" experiment with a general significance and scope that they probably did not deserve. He moves *very* quickly from the (admittedly fascinating and relevant) results of this study to a generalized argument for how background commitments and experience govern perception and therefore classification.

There's a place for this kind of phenomenon in understanding scientific classification, but it makes up a small part of a large and varied enterprise. To get a better sense of the general character of that enterprise, we return to Poincaré's discussion of experiment. He observes there that the act of classifying two experimental trials as two runs of the same experiment is an act of analogy. The specific circumstances that produce an experimental result are particular to a certain time and place, "circumstances . . . which . . . will never reproduce themselves all at once" (Poincaré 1946, 128). The circumstances are going to have to be close enough, and whether they are is a determination that the investigator will have to make. And, as every investigator knows, even maximally similar experimental circumstances are not sufficient to guarantee that she will feel compelled to count to two trials as two runs of the same experiment. Weird stuff happens all the time. The oil-drop notebooks of Robert Millikan, carefully studied by multiple historians and historically oriented scientists, are an important window into the acts of judgment by which investigators distinguish between experimental trials that can be used as part of a data set that provides us with the information that the experiment was designed to elicit, on the one hand, and trails that were "a failed run—*or, effectively, no run at all*" in Gerald Holton's words (1978, 209; his emphasis)—on the other hand. Sometimes it just doesn't work, and there's no apparent reason why. Thus, the classification of two experimental trials as two runs of the same experiment can only be made through analogy; individual runs bear enough relevant similarities to one another that they can be classified as individual instances of the same phenomenon.

The significant point for our purposes is that, often—probably far more often in scientific practice than in everyday life—classification is an act of the will, not a reflex of the mind. And those acts of the will must be based on judgments regarding what is relevantly similar to what. Now, of course, often times those judgments will be grounded in base perceptual capacities, or in the somewhat fancier perceptual capacities acquired through the practice of science; no doubt some of the subtle differences that motivated Millikan to discriminate against certain runs as "failed" were only detectable to him as a result of his experience with the oil-drop apparatus. Galison (1987) shares a lovely vignette involving Millikan's student, Carl Anderson, whose experiments with some of the first bubble chambers eventually led to the discovery of the positron in 1932. His research involved the examination of photographs of particle showers, which to his untrained eye presented an incomprehensible amalgamation of dots and tracks. But after several years of working with the photographs, he was able to distinguish certain kinds of particles from others, and to distinguish "good" (informative) photographs from "bad" (uninformative) ones. So the point is not that perception plays no role in scientific classification;

it clearly does. But *when* it does, it is typically part of a more comprehensive and deliberate approach to classification which includes perceptual experience within a larger corpus of considerations upon which a judgment as to how to classify an object will eventually rest—considerations keyed to a heterogeneous and mostly uncoordinated variety of goals associated with inquiry.

Together, this corpus of considerations leads us to our judgments regarding whether two objects are sufficiently analogous such as to be considered two members of the same class. This is the sense in which Poincaré saw any act of classification as an exercise in analogy. When we reflexively classify an object as a member of the kind *human*, we do so by drawing an analogy between certain salient characteristics marked out by that kind, and characteristics of the object we are classifying. Now, granted, it is an innocent analogy, by which I mean that it is not driven by deliberative judgment. It is a cognitive reflex. But it is an analogy nonetheless. Part of expertise involves the development of such a cognitive reflex as a result of one's initial concentrated and deliberate efforts to determine how best to draw analogies between the objects of experience. In general, to classify X as a member of Y is to judge that there is an analogy between X and other members of Y that is *strong enough* to warrant the inclusion of X in Y. The fact that these analogies are often made reflexively does not undermine their status as analogies. Rather, it reflects the fact that some uses of analogy involve a negligible degree of deliberative reflection. In the typical case, I would not be asked to give reasons for why I classify my wife as a human being. However, *were* someone to ask, I would gladly provide whatever evidence I expected my interlocutor to find compelling—her morphology, her intelligible speech, our thrice successful reproduction. I would, that is, point to specific features of her that best make the case for why she should be classified as a human (or maybe a superhuman).

In the context of research, acts of classification are frequently far less innocent than any such everyday acts as judging whether something is a human. In particular, the normative constraints on what sorts of inquiry tend to result in concrete successes must be paramount. Thus, when confronting a research problem, a practitioner's approach to classifying it will be guided by her knowledge of the kinds of problems she knows how to solve. Kuhn acknowledges this phenomenon directly in two places. First, when he describes the student's experience with problem sets in textbooks: "The student discovers a way to see his problem as like a problem he has already encountered. Once that likeness or analogy has been seen, only manipulative difficulties remain" (Kuhn 1974, 470). Second, in his appeal to the Galileo-Bernoulli lineage as of piece with the problem set example, in which Galileo "learned to see" an inclined plane "as like the pendulum with a point-mass for a bob" (Kuhn 1974, 470). But the latter example downplays a central aspect of the process—namely, that the way in which Galileo "learned to see" the inclined plane just so happened to be a way that allowed him to frame the inclined plane in terms of a problem he already knew how to solve (i.e., the pendulum). A ball rolling down an inclined plane is not begging to be seen as a pendulum with a point-mass for a bob. For purely visual reasons, there are a million other things with which we would analogize the ball-plane set-up before getting to the pendulum. More than a million. Heck, there is just no good reason to draw this analogy . . . *other than* that Galileo had already worked out in 1602 the fact that a pendulum's period was affected only by the length of the bob and not its mass: "this probably suggested to him the linkage with free fall" (Drake 1978, 72–73). In other words, Kuhn's casual switching from "seeing likeness" to arriving at a particular problem set-up elides the fact that only occasionally will the classification of a research problem be primarily explicable in terms of its perceived similarity to another problem. Rather, and for reasons which by now should be familiar, intuitive, and compelling, the normal case will be one in which the act of classification is driven overwhelmingly by the desire to exploit the problem-solving powers of an existing problem set-up. Off all the things a research problem *could* be seen to resemble, it would seem that more than coincidence would be required to explain why research problems consistently "come to resemble" problems we already know how to solve.

That something more is the normative pressure exerted by the desire to solve problems. In the face of this constant pressure, practitioners—like the students tackling problem sets—reach for one of the ready-made problem set-ups they have available to them, looking for ways to represent the problem in front of them as a version of the kind of problem they already know how to solve. This can be a relatively straightforward endeavor, as with the problem sets, where the problems are deliberately chosen to be easily conformable to a certain problem structure. And there are lots of additional cues—sample problems and such—that encourage the student to keep plowing ahead despite a string of failed attempts. But actual research presents the practitioner with a fundamentally different challenge. Here, when the application of an established problem-solving set-up is not straightforward, she must broaden her perspective with regard to the features that define the problem-kind exemplified by that set-up.

Darwin's argument from domestication in the *Origin* offers us a clear and well-studied instance of this process. Darwin's contemporaries accepted the empirical fact that a breeder could induce certain kinds of changes to his flock through selective breeding, allowing only the animals with his favored traits to reproduce. To show how adaptations arise in the wild, Darwin suggests an analogy between, on the one hand, the way a breeder prevents certain of his flock from reproducing and, on the other hand, the way in which environmental conditions prevent certain environmentally ill-favored members of a population from reproducing. The mere fact that breeding was sufficient to induce changes in a population was not by itself able to convince naturalists that adaptive transmutation was possible. Nor did Darwin assert that there was a single mechanism that is realized in one way in domestic settings and another way in wild populations. Rather, the argument rests on an analogy between the two contexts, one which invited readers to treat breeders and natural environmental conditions as two instances of a higher-level property.

Of course, this same kind of challenge arises in the context of students' problem sets as well. Remember our old friend from a few chapters back, the explosion? How is it that we come to frame an explosion as a time-reversed perfectly inelastic collision? Beginning physics students become well-practiced in setting up problems involving perfectly inelastic collisions, because their relative simplicity makes them a good system upon which to build up one's problem-solving intuitions. Now, I personally don't see any similarities between a collision and an explosion. To me they are opposites. But my introductory physics tells me that the category *perfectly inelastic collision* does not necessarily exclude explosions, because all that matters to a collision's perfect inelasticity is that the two objects have the same final velocity. Unlike many other collisions, and unlike many other final velocities, an explosion is a variety of collision in which the final velocity occurs at the beginning of the collision rather than the end; it is "time-reversed." However, since the direction of time makes no difference to how we analyze the forces in this scenario, I can still use my trusty "perfectly inelastic collision" problem set-up to solve the problem. In this way, I have been able to extend the problem-solving scope of that set-up by creatively reframing something that is pretty clearly not a collision in such a way that its differences with ordinary collisions are immaterial. From a problem-solving perspective, *perfectly inelastic collision* turns out to be a higher-level category that includes both time-reversed and standard varieties.

The transition to such a higher-level category is an instance of what I referred to earlier as a *stabilizing strategy*: we need to make a problem solvable, and so we create a version of it that represents it as analogous to a kind of problem we already know how to solve. The search for a means by which to bring a new problem under the umbrella of an existing problem-kind is an inevitable byproduct of the pressure to solve problems. Once a problem-kind has been exemplified, practitioners become intent on deploying it in as many contexts as possible. They thus begin to probe the space of puzzles—both solved and unsolved—with two principal goals: (1) to see whether there is an unsolved puzzle which could be made solvable by framing it as a version of the recently minted new kind of problem; (2) to see whether we gain valuable new insight by framing solved problems as versions of the new problem-kind. Any effort to apply the new set-up—that is, to frame some phenomenon as an instance of the new problem-kind—will involve a series of choices regarding how exactly to bring the set-up and the phenomenon into correspondence. We regard certain aspects of the frame (such as the final velocities of the colliding objects) as particularly salient and important; others (such as the direction of time) we consider merely incidental. The more successes practitioners tally up in their campaign to colonize idea space with the new problem-kind, the more they begin to cling to it as an indispensable problem-solving tool, and the more entrenched it becomes. Consequently, they experience a corresponding relaxation in pressure to innovate. As long as the depth and breadth of insight continue to be worth the effort required to represent a puzzle as part of the new problem-kind, practitioners will have a strong incentive to stay within that framework.

The strategic and artful application of a given problem-solving set-up ultimately results in the emergence of higher-level groups that include the original problem-kind as one species of a broader genus.[1] The emergence of such groups corresponds to one variety

---

[1] It is this process that results in the form of explanatory unification articulated by Kitcher (1989, 432): in an effort to expand the range of applicability of a given scientific explanation, practitioners adopt a more abstract perspective on the explanation's "non-logical expressions" (theoretical terms and such), resulting in "schematic sentences" that together form the "schematic argument" from which the explanation for a given phenomenon is derived.

of fruitfulness—namely, the unpredictable proliferation of diverse yet related problem-solving tools. The emergence of these broad and diverse groups is a direct consequence of the pressure to solve problems, and the associated commitment to the broadest possible application of a given solvable problem set-up. The farther practitioners extend the scope of a given set-up, the more abstractly they are forced to interpret the components of that set-up—at a certain level of abstraction, explosions are *also* a kind of collision; at a certain level of abstraction, adaptive transmutation is *also* a kind of breeding. Out of these forced abstractions emerge the higher-level problem-kinds that I have called *families of solvable problems*. On my account, fruitfulness is the propensity to generate such families. Fundamentally, this propensity is driven by the need for a general capacity to structure problems in a solvable form, a capacity which is furnished by exemplars.

6.3. Fruitful Metaphors

As we have observed, *Structure* conceived of these paradigm shifts overwhelmingly in terms of shifts in vision. Kuhn, it seems, felt that this gave him the ability to explain why practitioners' perceptual experiences were different after a paradigm shift. Beginning in the 1970s, however, Kuhn came increasingly to think of these shifts—that is, of "revolutions"—as the result of developments in the referential behavior of natural kind terms.[2] I interpret this development in Kuhn's thought as essentially a major refinement of his roughly articulated views in *Structure*. He was open regarding the fact that he saw the Kripke-Putnam causal theory of reference (CTR) as a "great advance" and its utility as a framework for elaborating his own views is evident from the mid-1970s onward

---

[2] Of course, elements of this perspective are already present in *Structure*'s discussion of Newtonian mass and Einsteinian mass in Chapter 9.

(Kuhn 1979, 411). Of course, Kuhn understood the history of science too well to be lured into thinking that the nature of reference-fixing in science was generally as straightforward and well-behaved as the early versions of CTR suggested (see, e.g., Kuhn 2000; also 1979, 412–414). But its thematic effect on his thinking about paradigm shifts—which, following the philosophers, he would regrettably refer to as "theory change"—was nevertheless substantial. The centrality of perceptual experience as an explanatory center of gravity drops out of Kuhn's work almost entirely thereafter. It survives as *one* of the causal factors that can precipitate scientific revolutions, now understood as changes in the referential behavior of natural kind terms (Wray 2011, 27). That perceptual experience would retain this role in some form in Kuhn's framework was unavoidable. For one thing, *Structure* places almost the entire explanatory burden of the precipitation of "crisis" on violations of the practitioner's intuitive expectations for how nature should work, expectations that have, through scientific training, become part of the lens through which they experience the world (Kuhn 1962/2012, 64–65). Second, Kuhn literally says that only empirical anomalies (as opposed to logical or philosophical objections) can precipitate a crisis (Kuhn 1962), a view he reiterated even after his turn to language was in full swing: "alterations in the way scientific terms attach to nature . . . come about in response to pressures generated by observation or experiment" (Kuhn 1979, 416). From this vantage point, Kuhn's mature view is essentially a heavily modified descendant of that expressed in *Structure*, one which preserves in some form the causal significance of perceptual experience as a repository of tacit knowledge, but which decentralizes this element in a way that brought his general framework into more direct contact with the dominant tenor of debates in contemporaneous philosophy of science.

At the center of the modified, mature expression of his view, articulated throughout the 1980s and 1990s, scientific revolutions occur when some discovery requires that the extensions of distinct natural kind terms must overlap, thus violating the "no-overlap" principle governing the use of such terms. For example, our physical theory holds that the terms "electron" and "proton" refer to distinct natural kinds, and that, given that they are distinct natural kinds, we could not discover that, say, electrons were actually a kind of proton; the extensions of those terms do not—*cannot*—overlap. If, contrary to all expectation, scientists discovered that the proton *was* a kind of electron, "they cannot just enrich the set of category terms but must instead redesign a part of the taxonomy" (Kuhn 2000, 92).[3] The "no-overlap" principle has been violated, and a new worldview must be elaborated, one built out of taxonomically well-behaved categories that dutifully obey the no-overlap principle governing natural kind terms.

As Kuhn helpfully indicated in a number of his later articles, the no-overlap principle governing natural kind terms was adjacent to the principle of noncontradiction, in how it specifies that to be a member of one natural kind necessarily precludes membership in another kind. But he also acknowledged the existence of "recognized ways of bracketing the rule"—devices such as *metaphor*, for instance, where the speaker and listener agree not to assess a statement according to the strictures on literal truth (Kuhn 2000, 100). Indeed, some years before, he had gone so far as to state that metaphor was a "higher-level version of the process by which ostension enters into the establishment of reference for natural-kind terms" (Kuhn 1979/2000, 201). Commenting on the metaphor, "War is a game," he observed that

> [t]he actual juxtaposition of a series of exemplary games highlights features which permit the term 'game' to be applied to nature. The metaphorical juxtaposition of the terms 'game' and 'war' highlights other features, ones whose salience had to be

---

[3] Quoted in Wray (2011, 25–26).

reached in order that actual games and wars could constitute separate natural families. (Kuhn 1979/2000, 201)

Around this time, he articulated in broad outline his revised conception of scientific revolutions, in which he credited "changes in
model, metaphor, or analogy" as "probably the most consequential" features of scientific change: "Violation or distortion of a
previously unproblematic scientific language is the touchstone of
revolutionary change" (Kuhn 2000, 30–32). The original title of that
essay was "From Revolutions to Salient Features" (Kuhn 2000, 13).

Clearly, the role of metaphor in science appeared on Kuhn's radar
shortly after writing *Structure* and stayed there for a very long time.
Yet, despite his awareness of metaphor's close relationship with his
mature, "no-overlap"-centered account of scientific revolutions, he
never explored the specific mechanics of metaphor with any depth.
This was, I believe, a serious oversight. Let us look at four reasons
for this belief, listed in order of increasing degrees of lamentability.

The first reason is that, plainly, there is no meaningful difference
between his mature theory of revolutionary change and the general
theory of metaphor. Nelson Goodman provides an able account of
the canonical understanding, one which mirrors that of Max Black
(whom Kuhn cites approvingly [1979/2000, 197]) and many subsequent theorists.[4] Goodman's account of metaphor "begins with the
recognition that a label functions not in isolation but as belonging
to a family" (Goodman 1968, 71), a thoroughly Kuhninan senti-
ment that resonates particularly strongly with later expressions of
his taxonomic theory of meaning holism (see, e.g., Kuhn 2000).
The use of the label *dog*, for example, involves more than just reference to a certain kind of creature. Rather, it invokes a whole set
of notions—many of them tacit—bound together through their association with dogs. This family of notions has a certain organiza-
tion or structure to it, ordered according to the degree of salience

---

[4] See Camp (2006, 2009, 2019a); Bowdle and Gentner (2005).

assigned by context. When I use the label *dog* literally, as in "Look at that dog," the notions that are called forth correspond to the core properties which we associate with the natural kind *dog*—the bark, the muzzle, the pronounced canines, paws, waggly tail, and so on— the stable cluster of properties that we use in practice to distinguish dogs from non-dogs (Slater 2015). Goodman called this structured set of notions a "schema." It corresponds to the notion of a frame described in the previous chapter (*sensu* Camp 2019a).

In the same way that all dogs fall within the extension of the label *dog*, Goodman pointed out that each notion in a frame also comes with *its* own extension; "paws" refers to paws, "teeth" to teeth, and so forth. According to his view, to use a term metaphorically is to use it in territory that lies beyond the native range of extension of the notions that constitute its frame. For instance, when we apply the label *dog* to a certain man, as in, "Jerry's a dog," we transfer the frame connected to *dog* into an "alien realm," one whose occupants lack properties that fall within the extension of "paws," "waggly tail," and so on. In Kuhn's terms, we have violated the "no-overlap" principle; realms are "alien" to one another just in case their literal extensions do not overlap. In this way, the use of metaphor is a "violation or distortion of previously unproblematic language." Given his mature view's indistinguishably close association with metaphor, I rate his failure to make use of the available resources on metaphor "somewhat lamentable."

The second reason why this qualifies as a serious oversight derives from the fact that Kuhn explicitly associates the Copernican revolution—his cherished exemplar of revolutionary change—with the use of metaphor:

> Metaphor plays an essential role in establishing links between scientific language and the world. Those links are not, however, given once and for all. Theory change, in particular, is accompanied by a change in some of the relevant metaphors and in the corresponding parts of the network of similarities through

which terms attach to nature. The earth was like Mars (and was thus a planet) after Copernicus, but the two were in different natural families before. (Kuhn 1979/2000, 203–204)

The sense of metaphor being invoked here aligns neatly with the canonical understanding of metaphor given earlier. From the Ptolemaic perspective, the "relevant metaphor" would be "Earth is a planet." The literal extension of the label *planet* included Mercury, Venus, Mars, Jupiter, and Saturn. These were the celestial bodies whose behavior differed from that of the vast majority of others in a particularly significant way: they "wandered" around the sky. Now since, according to the Ptolemaics, the earth is stationary, it lies beyond the native range of the extension of the term *planet*. Thus, to label the earth as a planet would be to transfer the frame connected to *planet* into an alien realm—the earth—where the family of notions associated with *planet* do not literally hold.

Appropriately, this episode can be couched in terms of Kuhn's mature theory of revolutionary change.[5] From the perspective of Ptolemaic astronomy, Copernicus's *De Revolutionibus* engages in an egregious violation of the no-overlap principle. In the Ptolemaic framework, celestial bodies divide into two distinct natural kinds: stars, which maintain fixed positions; and planets, which don't. Mars, as Kuhn notes, is a wanderer, a planet. The earth is stationary, so definitely not a planet. The Ptolemaic natural kind *planet* excluded stationary celestial bodies, the earth among them. By framing the earth as a planet, Copernicus caused a category with stationary members to overlap with a category whose members were in motion.

It's worth looking at the Copernican episode in a little bit more detail, because the heart of the transition to heliocentricity bears a strong resemblance to the deliberate use of metaphor, and possesses core features that are not at all well-captured by Kuhn's

---

[5] See Wray (2011, chapter 2) for discussion.

account of how revolutionary transitions arise. For this reason, I rate his failure to develop his thoughts on metaphor here "quite lamentable." With respect to Copernicus's own motivations, as far as we can tell they were centered on his prejudice against planetary models that were not physically realizable, a prejudice to which other Ptolemaics seem to have been largely immune (Swerdlow and Neugebaur 1984, 60). Copernicus was able to achieve a physically possible planetary model by fixing the sun as a stationary center around which other bodies, including the earth, were in orbit.

If this is correct, the beginnings of this transition differ from Kuhn's model in a number of important respects. For one thing, the development of the heliocentric theory does not proceed on the platform of an incumbent theory mired in crisis. The demand that a planetary model be physically realizable was one of a number of constraints that Copernicus had set for himself (such as the return to genuinely uniform circular motion) (Swerdlow 2012, 372). While it may not have been unique to Copernicus (the Maragha school had also worried about this), it was not a problem that astronomers as a community felt pressured to solve; in Swerdlow's words, "he was the only person of his age who made an original contribution to astronomy at all" (Swerdlow 2012, 371–372). Relatedly, the failure of geostatic models to accommodate that constraint cannot be meaningfully categorized as an "anomaly" in the Kuhnian sense: since no part of the Ptolemaic paradigm required that planetary models be physically possible, there was no generally appreciated reason to fuss over the difficulties this constraint presented for geostasis. Lastly, and again relatedly, it is not at all clear what role aberrant perceptual experience could have played in inducing Copernicus to "see" the earth as a planet. Indeed, what does seem clear is that framing the earth as a planet was a deliberate choice, one which sets him apart from every other astronomer of his age, "as far as is known" (Swerdlow and Neugebaur 1984, 59). In other words, there is no plausible sense in which Copernicus "learned to see" the earth as a planet (I mean, maybe at some point later in life . . .). The opening salvo of the revolution was a conscious effort by Copernicus to frame the earth as a planet to satisfy a particular research interest of his while preserving in broad outline the problem-solving toolkit of Ptolemaic astronomy. It is for his fidelity to that toolkit that he is regarded as the last great Ptolemaic astronomer.

We now come to the "deeply lamentable" part of the saga. Is there anything more Kuhnian than the idea that a wide and diverse range of unexpected discoveries and novel perceptual experiences result from a practitioner's shift in perspective? No, there isn't. It is the most Kuhnian idea ever. I think that is why Kuhn was so attracted to it. Indeed, he makes a very compelling case for understanding the revolutionary status of *De Revoliutionibus* precisely in these terms:

The history of astronomy provides many other examples of paradigm-induced changes in scientific perception, some of them even less equivocal. Can it conceivably be an accident, for example, that Western astronomers first saw change in the previously immutable heavens during the half-century after Copernicus' new paradigm was first proposed [i.e., before the telescope]? The Chinese, whose cosmological beliefs did not preclude celestial change, had recorded the appearance of many new stars in the heavens at a much earlier date. Also, even without the aid of a telescope, the Chinese had systematically recorded the appearance of sunspots centuries before these were seen by Galileo and his contemporaries. Nor were sunspots and a new star the only examples of celestial change to emerge in the heavens of Western astronomy immediately after Copernicus. Using traditional instruments, some as simple as a piece of thread, late sixteenth-century astronomers repeatedly discovered that comets wandered at will through the space previously reserved for the immutable planets and stars. The very ease and rapidity with which astronomers saw new things when looking at old objects with old instruments may make us wish to say that, after Copernicus, astronomers lived in a different world. In any case, their research responded as though that were the case. (Kuhn 1962/2012, 116–117)

Kuhn provides no account of the mechanism by which these shifts in salience occur. But the well-established cognitive behavior of metaphor does. And it corresponds damn near perfectly to the most canonical form of fruitfulness—namely, that associated with unforeseeable discoveries that seem to flow naturally, almost inevitably, from an event. In the remainder of this section, I describe how the normative status of a given problem-solving set-up leads to the development of metaphors that bring about the shifts in salience that Kuhn always maintained lay at the center of revolutionary change.

Interestingly, as Tappenden has shown, this dynamic was also a central preoccupation of Frege's epistemology of mathematics, as reflected in this characteristic passage:

The more fruitful type of definition is a matter of drawing boundary lines that were not previously given at all. What we shall be able to infer from it, cannot be inspected in advance; here we are not simply taking out of the box again what we have just put into it. The conclusions we draw from it extend our knowledge . . . and yet they can be proven purely analytically and are thus analytic. (Frege 1980: 100–101)

Frege saw the development of mathematical knowledge as largely dependent on choosing fruitful concepts and definitions with which to frame mathematical problems. For him, a fruitful definition was one that could "extend our knowledge" by "drawing new boundary lines" out of which new mathematical insights emerge. Indeed, "it is because of fruitful concepts that new knowledge is potentially available from analytic judgments" (Tappenden 1995, 428).

For Frege, the fruitfulness of a definition was not a matter of chance. On his view, the extensions of our knowledge that flow from fruitful definitions "are, in fact, contained in the definitions, but like a plant in a seed, not like a beam in a house" (Frege 1980, 101). Unlike the beams of a house, which are constitutive of the house, the plant "in" the seed requires certain favorable conditions to be "brought out." Although the seed may offer no indication of harboring a plant, it nevertheless "contains" the plant in the sense that, if placed in the appropriate setting, a plant naturally follows from it. So it is with fruitful definitions. There are things "inside" certain definitions and concepts that only reveal themselves under certain conditions.

Dummett (1973) complained that this "plant" metaphor was "no great help."[6] On the contrary, I think it is extremely revealing, for it brings Frege's thoughts on fruitfulness into direct contact with the study of metaphor, particularly through Goodman's work. In this section, I will argue that the capacity for fruitfulness which Frege valued in some definitions—and which is generally valued across the sciences and mathematics—is routinely accessed in rational inquiry through the metaphorical use of problem-solving frames developed in one area and then applied to another. I then show how the elicitation of this capacity through metaphor bears directly on Kuhn's views concerning scientific revolutions and "no-overlap" principle.

In metaphorical contexts, the mind's reflex seems to be to prioritize the continued stability of the structured set of constituent notions. To maintain stability in alien territory, we mentally restructure the alien realm to mirror the structure of the invading frame. Fascinatingly, this restructuring process appears to be in large part automatic.[7] While "[t]he choice of territory for invasion is arbitrary," writes Goodman, the specific features of the alien realm

---

[6] Dummett (1973); cited in Tappenden (unpublished; 2018, 13).
[7] As described by Bowdle and Gentner (2005), along with many other linguists.

that are affected by the restructuring are "very largely determinate." This is because the literal use of a frame is a matter of its habitual application. When the frame is used to generate characterizations in a nontraditional context, the pattern of that habitual application does not simply evaporate. It remains the dominant force in determining how that frame will affect characterizations of the alien realm. As Goodman (1968, 74) observes, "Antecedent practice *channels* the application. . . . [B]y thus carrying with it a reorientation of a whole network of labels does a metaphor *give clues for its own development and elaboration*."[8] Richard Moran eloquently captures this phenomenon when he notes:

> Part of the dangerous power of a strong metaphor is its control over one's thinking at a level beneath that of deliberation or volition. In the mind of the hearer an image is produced that is not chosen or willed. The metaphorical assertion brings one to see something familiar through this image, framed by it, and this "seeing" persists concurrently with one's original sense of the dissimilarity of the two things here being brought together. (Moran 1989, 90)

Let's draw out the connection between Frege's and Goodman's views by indulging a little further in the biological metaphors to which they both appeal. Beyond his description of metaphor as the "invasion of alien territory," Goodman also observed that a really successful metaphor "requires a combination of novelty with fitness" (Goodman 1968, 79). Whether purposeful or not, Goodman has nicely captured the process by which an evolutionarily successful invasion occurs. This process begins when a species arrives in a territory that (1) differs significantly from its native habitat, but (2) nevertheless provides conditions that are hospitable to that

---

[8] Goodman 1968, 72; (my emphasis).

species occupying that territory. In linguistic contexts, this would be an instance of an apt metaphor.

Next consider Goodman's observation that the "choice of territory for invasion is arbitrary," but the nature of the change affected in the realm is "very largely determinate." The most widely accepted theory of metaphor among contemporary analysts is *structural mapping* theory. A term used literally is in its "native environment." Hearers reflexively use the term's literal meaning to guide their search for a plausible interpretation. When the term is used literally, the guidance provided by the term's literal meaning will produce search results that fall within the term's literal extension. But the literal meaning *also* guides the interpretive search when a term is used figuratively, in "alien territory"; the semantic reflexes that guide a term's interpretation will prevail in any "environment." When these reflexes guide a search of alien semantic territory, they restructure the territory in a way that permits a successful "mapping"—that is, in a way that permits the interpretive search to conclude successfully. Thus, the metaphor "carr[ies] with it a reorientation of a whole network of labels" in the sense that the rules that guide interpretation in native settings continue to guide interpretation in nonnative settings.

I believe that it is a version of this phenomenon which Frege sees in the capacity for a definition to be fruitful, to "give clues for its own development and elaboration," in Goodman's words. As Tappenden has shown, "the idea of the 'organic' was seen as inherently linked to creativity by people close to Frege, establishing a background conceptual connection between 'fruitfulness' and organic structure."[9] Like organisms, linguistic or cognitive frames—concepts, terms, and definitions—have native environments, contexts that tend to bring out the sorts of features that we characteristically associate with them and which constitute their "normal" or literal functioning. But different contexts may elicit very different features

[9] Tappenden (unpublished; 2018, 18).

or shift the salience balance in unforeseeable ways. Applying a frame in a nonnative context thus has the power to unlock latent potential for "development and elaboration" just as can occur when placing a species in a nonnative environment. The unlocking of latent potential is the final variety of fruitfulness in science and mathematics that I will consider.

There is a pronounced trend across the history of science and mathematics of practitioners generating entirely novel families of solvable problems through the application of an established frame in a nonnative context. It is precisely this dynamic that Frege has in mind when he refers to the "scientifically fruitful" consequences that follow when "totally new boundary lines are drawn by . . . definitions"—"Here too, we use old concepts to construct new ones, but in so doing we combine the old ones together in a variety of ways."[10] A framing device such as a metaphor attempts to "draw new boundary lines." The literal use of the frame *dog* does not include men within the boundaries of its extension, but the metaphor "Men are dogs" redraws the frame's boundary lines to include men. Once men are implicated within the extension of *dog*, our conceptualization of men is hijacked by the salience norms that characteristically guide interpretation when we predicate of something that it is a dog. The influence of the salience norms governing our conceptualization of men is temporarily suspended or diminished, while unnoticed or underappreciated dog-like features of men are brought to the fore.

This process, by whichever of the many names it has been called—violating the "no-overlap" principle, "drawing new boundaries," using metaphors, representing-as, reframing—is as central to the process of discovery in science and mathematics as it is inevitable. Earlier I argued that the development of a research frame, a set of constraints for guiding research, is an object of intense demand in inquiry and is not easily given up. Due to its

---

[10] Frege, "Boole's Logical Calculus," in Frege 1979. Quoted in Tappenden (1995, 431).

inquiry-facilitating capacities, practitioners are motivated to apply a particular frame in as many instances as possible, using it to generate characterizations of phenomenon after phenomenon.

In the most straightforward cases, a new phenomenon will maximally resemble the exemplar upon which a frame is grounded—as, for example, when deciding whether two morphologically indistinguishable organisms are members of the same species as the "type-specimen" that is used to guide the identification of other members. We include these conspecifics in the same class as the exemplar through an analogy between the exemplar's salient or characteristic features, on the one hand, and the salient features of the other creatures, on the other. Applying the frame in this way allows the new phenomenon to stay in its "native psychological environment," in that it resembles as closely as possible the features that the frame was designed to capture.

But not all uses of a frame are this straightforward. Just as a species will, if it continues to roam, encounter new kinds of environmental challenges, the attempt to apply a frame to a greater and greater range of phenomena will inevitably lead to incongruities between the conditions required for literal application, on the one hand, and the salient features of the subject that the frame is being used to characterize, on the other. Practitioners will nevertheless push on, because they are loathe to relinquish the problem-solving power of particularly useful frames. They are thus forced to take a more figurative, more abstract, or in general more liberal perspective on how a subject might legitimately fall within the scope of the frame. Like a species in an alien environment, the use of a frame in novel contexts will require a style of application that simply did not arise in its native environment.

As the style of application shifts to a less literal, more metaphorical approach, more research problems come to be characterized by the frame. Eventually the frame is used to characterize a problem in a way that produces unexpected and unforeseeable results. Because practitioners use the frame to guide their approach to problems—to ask certain questions rather than others, to treat certain features as particularly salient or important—they end up investigating features of experience that they would otherwise never have seen or considered.

Now because "antecedent practice channels application," the results of applying the frame in novel contexts are significantly constrained and "very largely determinate"; were this not so, the understanding of metaphors and analogies would not be so uniform (see next chapter). But the constraints governing the application of the frame produce different results in different contexts. When applied to one kind of problem, the frame's reorganization of the subject terrain generates a characterization that highlights certain kinds of features. When applied to another kind of problem, the reorganization effected by the frame highlights other kinds of features. While saying, "He's a dog" in reference to someone's husband is intended to highlight a certain male tendency to a kind of opportunistic and hedonistic disloyalty, saying "He's a dog" in reference to someone's cat might be intended to highlight the cat's uncharacteristic *loyalty*. As Bowdle and Gentner (2005, 197) observe, "these metaphors suggest different meanings . . . because metaphors invite alignments among different systems of predicates." Emily Grosholz vividly describes this dynamic in the context of mathematics:

> Important problems often reorganize and extend mathematics via the processes by which they are solved. All of them work by indirection, and establish new correlations that transcend or traverse the boundaries of domains as they had been settled up to that point. This organization and extension is unprecedented, but once established the determinacy of the things correlated renders it determinate as well and far from subjective. Emerging conditions make new things and new alignments possible. (Grosholz 2007, 191)

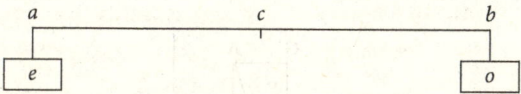

Figure 6.1 Euclidean representation of a balance.

In the same way that a species' developmental constraints give rise to new lineages under novel environmental conditions, the same frame will give rise to new families of solvable problems when applied in research contexts that lie beyond its native boundaries. In both instances, a new developmental pathway emerges from the constraints imposed by an established structure built for problem solving under other conditions.

Illustrations of this process are rampant across the history of science. I provide two here: Galileo's little balance diagram (Fig. 6.1) from *De Motu* and the diagram for his proof of the mean speed theorem in *Discorsi*. This drawing of a balance is meant to illustrate Galileo's early equilibrium theory of motion, of which I'll kindly spare the details, and which he in any case abandoned around the turn of the seventeenth century (Machamer 1998, 58). While it is meant to be a drawing of a balance, the particular style of representation is designed to highlight features of a balance that correspond very closely to a Euclidean line. Not all depictions of a balance will bring these features out; a more realistic depiction could not succeed in this effort, for instance. This specific way of representing a balance suggests that certain of the tools that are used to solve problems in geometry might profitably be applied to problems of motion like those exemplified in the operation of the balance.

Similar considerations apply to his diagrammatic proof (Fig. 6.2) of the mean speed theorem—that is, that an object in uniform motion will, in a given time, traverse the same space as an object uniformly accelerated from rest (along AE) whose final speed is twice that of the first object. Now, I have emphasized throughout this book that the guidance—often through base perception—of

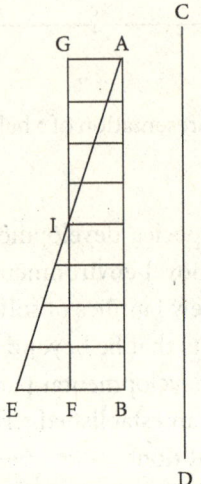

Figure 6.2 Galileo's proof of the mean speed theorem, from *Opere* 8: 208.

previous problem-solutions as envisioned by Kuhn after *Structure* is at best a special case of the phenomenon that drives the development of families of solvable problems. Even if we grant that some aspect of this diagram's resemblance to an inclined plane is an instance of the perceptual special case Kuhn describes, what of others? What about, for instance, the line *AB*? This is Galileo's representation of all the moments of time between the beginning of the object's motion and its final velocity. I see no reason to think that either Galileo or any of his readers (including me) "came to see" time in terms of a vertical line *AB*. It's just a way of diagrammatically representing moments of time. Nor can I see what explanatory gain might accrue by supposing that anyone did. The proof is compelling even if we think of *AB* as a metaphor-like representation of time. It shows how the problem of uniformly accelerated motion in free fall could be represented as a problem in Euclidean geometry—that is, a kind of problem everyone knew how to solve.

In this way, Galileo provides not only a proof of a theorem with a direct connection to the science of motion, but a demonstration of a representational vehicle for the problems of motion which makes those problems solvable. It is a solved problem that functions as a *sample* of problem solving.

And it was interpreted as such. The development of mechanics through the seventeenth century is a testament to the inherent generative power of frames, as the Euclidean framework perpetually suggested new questions, new tools of investigation, and further extensions of the frame to other types of phenomena and other quantities. Before the end of the century, Newton would carry the frame further than anyone in Galileo's time could have imagined. When combined with his inertial physics, it gave practitioners the means with which to explore a vast range of phenomena, and with which to geometrically express both the magnitude and direction of a force, as in the *Principia*'s Proposition 6, Theorem 5 (see Fig. 6.3).

Damerow et al.'s (1991) encapsulation of this process, and its role in the development of classical mechanics (quoted here at length), does able justice to the generative power of frames that are "flexible enough to cope with a wide range of experience, [yet] rigid enough occasionally to display inconsistencies when separate legitimate applications of a concept lead to extensions of meaning which turn out to be incompatible":

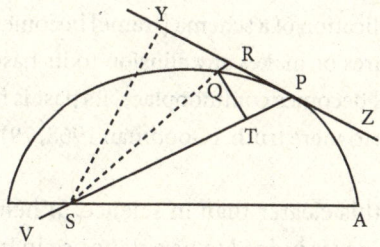

Figure 6.3 QR is the force required to pull a planet away from its inertial path.

The conceptual development embodied in the transition to classical mechanics cannot be identified with any particular way station and is not to be found in any particular text. It is a process which begins with such figures as Descartes and Galileo and takes shape with the generation of their successors. These disciples or even adversaries read the old problems and arguments from the point of view of their new solutions, thus establishing classical mechanics, because their point of departure was now the concepts as they are implicitly defined within the derivations of the theorems, e.g., the law of free fall. Thus, while for the first discoverer, the law of free fall is achieved by applying and modifying an independently grounded, pre-existing conceptual system, for his disciples it is the law of fall that canonically defines key concepts in a new conceptual system. The very same reading of these theorems that establishes classical mechanics also obliterates the traces of its real historical genesis because the original problems and the concepts involved are now understood within a very different theoretical and semantic framework. (Damerow et al. 1991, 5)

In related terms, Goodman makes a final, provocative observation as he concludes the section on metaphor in *Languages of Art*:

Since metaphor depends upon such transient factors as novelty and interest, its mortality is understandable. With repetition, a transferred application of a schema [frame] becomes routine, and no longer requires or makes any allusion to its base application. What was novel becomes commonplace, its past is forgotten, and metaphor fades to mere truth. (Goodman 1968, 79)

In no context is this clearer than in science. When a metaphor or other framing device is judged to be apt, its use in inquiry becomes not only tolerated but encouraged and routine. What was for one generation of practitioners a revolutionary shift in perspective is for later generations just business as usual. Once again, Frege describes this dynamic in organic terms:

> The natural course of events seems to be as follows: what was originally saturated with thought hardens into a mechanism which partly relieves the scientist from having to think. Similarly, in playing music, a series of processes which were originally conscious must have become unconscious and mechanical so that the artist, unburdened of these things, can put his heart into playing. I should like to compare this to the process of lignification. Where a tree [Baum] lives and grows it must be soft and succulent. But if what was succulent did not in time turn to wood, the tree could not reach a significant height. On the other hand, when all that was green turns to wood, the tree ceases to grow.[11]

Like all metaphors, repeated use of a research frame contributes to its conventionalization. As part of what they call "the career of metaphor," Bowdle and Gentner (2005) argue that as metaphors become conventionalized, they develop both literal and figurative meanings. And, indeed, Blank (1988) showed "that conventional metaphors are processed as quickly as literal sentences, whereas novel metaphors took significantly longer."[12] Likewise, through repeated use and extension as an organizing motif for an increasingly wide range of research domains, the frame that once began as a metaphorical cast of a particular research problem increasingly comes to be seen not just as useful vehicle for generating apt characterizations of nature, but as encoding a privileged perspective on what nature is like.

Readers of *Structure* will recall the way in which Kuhn was palpably enamored of this process, which he believed to be endemic to the nature of research and an essential ingredient in the

---

[11] Frege (1980, 33); quoted in Tappenden (2018; unpublished).
[12] Cited in Bowdle and Gentner (2005, 200).

provocation of the kind of confrontation with theory that leads to scientific revolutions. Although I am convinced that the "lignification" of a research frame is a significant determinant of the efficiency and scope with which practitioners apply it, to say that the frame typically becomes part of how practitioners "see the world" is probably not accurate. Unquestionably, it sometimes does; I think many of the cases that Kuhn adduces in support of this thesis are genuine instances of it. But it also very often does not. It does seem that practitioners are often far from being perceptually locked into a certain perspective on phenomena in the way that Kuhn would have us believe is generally the case. A general understanding of paradigm choice as a choice of metaphor can accommodate this Kuhnian staple as a special case via the tendency of metaphors to be conventionalized into literal truths. In the main, however, the compulsion to set up research problems in a specific way need only derive from the practitioner's commitment to the privileged efficacy of that particular approach to problem set-up. I explore the forms that privilege can take in Chapter 8.

6.4. Conclusion

Exemplars, as Kuhn said, "attract an enduring group of adherents away from competing modes of scientific activity." Practitioners form an attachment to the exemplars in their disciplines; they *cling* to them. Kuhn believed that the bonds of attachment were held together by the community's belief that they pretty much understand how the world works. But we've now looked at several reasons why that view is not plausible as a general account of why paradigms or research frameworks exert a normative influence on the practice of science or mathematics. This chapter has given us yet another reason to doubt it: analogical and metaphorical representations of research problems are unavailable for the ontologically committed. For them, exemplars exemplify some aspect of how the world is.

From this vantage point, to engage in nonliteral uses of the exemplar would be to abandon one's ontological commitments and thus drain the exemplar of its normative power. Without that normative power, there can be no normal science: there is no "enduring group of adherents," because there is no worldview for them to adhere to. Yet the development of analogy and metaphor is inevitable once a problem-solving exemplar has emerged.

Nonliteral forms of representation are the expected result if we assume that what exemplars exemplify is a particular problem-kind: the drive to solve problems compels practitioners to cling to ways of setting up problems that are known to make them solvable. Adopting a nonliteral—analogical or metaphorical—perspective on the features that are made salient by exemplars presents no obstacle to their continued use *as long as* exemplars do not demand any ontological commitment. And, if what exemplars exemplify is a way of structuring problems, they make relatively few demands on a practitioner's worldview. This is not to say that exemplars make *no* demands. In particular, as I've been arguing, exemplars prescribe a certain way of characterizing a research problem. By prescribing a certain set-up or characterization for a problem, they invite practitioners to think of or conceptualize that problem in a certain way.

The effects of these prescribed characterizations are like those of metaphor, because prescribed characterizations mandate the "transfer" of a problem to an "alien realm." When practitioners use an exemplar to guide their set-up of that problem, they force it into a particular frame, and we expect the cognitive effects of framing the problem in that way to proceed just as they do in more conventional uses of metaphor. In framing the balance as a Euclidean line, Galileo transferred a problem in statics to the alien realm of Euclidean geometry. That realm accentuated the salience of certain features of the balance, such as the magnitude of displacement of one side, or the angle traversed by a line as it is pulled down by a weight. These properties of the balance don't necessarily jump out at you when you look at it, but they are highly salient features of Euclidean set-ups. Framing the balance as a Euclidean system brings these features irresistibly to the fore.

Every seasoned practitioner is aware of the threat of certain inevitabilities that result from thinking of a phenomenon in a particular way. This is simply an occupational hazard: we must think of things in some way or other. To adopt an exemplar, or to classify a problem in a certain way, is to make a choice regarding what metaphor ought to be used to frame a research problem. The next chapter explores the species of warrant that accompanies such choices, and it argues that this species of warrant is the alternative variety of epistemic justification we set out to uncover.

7

The Aptness Alternative

*From the point of view of the physicist, a theory of matter
is a policy rather than a creed; its object is to connect or co-
ordinate apparently diverse phenomena, and above all to
suggest, stimulate and direct experiment. It ought to furnish
a compass which, if followed, will lead the observer further
and further into previously unexplored regions. Whether
these regions will be barren or fertile experience alone will
decide; but, at any rate, one who is guided in this way will
travel onward in a definite direction, and will not wander
aimlessly to and fro.*

—J. J. Thompson, *The Corpuscular Theory of*
Matter (1907)

A little while back, I attended a colloquium in our physics department on the LIGO detection of gravitational waves. Their detection constituted a novel successful prediction made by the theory
of general relativity, and it was hailed in the popular press as confirmation of Einstein's theory. And, indeed, my own default perspective was to view the LIGO experiment as, fundamentally, a test of
that theory. During the question-and-answer period, I asked how
many tests of general relativity were going to be enough before the
physics community was prepared to conclude that it was correct.
The resident cosmologist patiently responded, "The importance
of the detection of gravitational waves is not really about whether
Einstein's theory is correct. It is about what we now know we will be

Fruitfulness. Chris Haufe, Oxford University Press. © Oxford University Press 2024.
DOI: 10.1093/oso/9780197666395.003.0007

able to do in the future, about how far we'll be able to look back into the early universe, etc." In other words, this cosmologist viewed the formal relationship between general relativity and gravitational waves as more or less inconsequential. In retrospect, this should have been obvious to me. My question presupposed that the epistemic significance of the LIGO experiment lay in evaluating the truth of Einstein's theory. Despite how important the truth of a theory may be to *believing* or *accepting* it, though, it is not rationally incumbent upon practitioners to value a theory's truth in a research context. This is because there is a lot you can do with a theory that does not involve accepting its truth into your heart.

What exactly do we want out of an account of epistemic justification, anyway? What are we hoping for? Speaking for myself alone, I guess I'd like it to be able to give me a sense of why decisions made on a certain basis tend to "pay off." As any congressional lobbyist will tell you, payoffs come in different forms. The epistemology of the last one hundred or so years has demanded payment in the form of truth. Now, I'm as fond of truths as anyone. The truth tells it like it is. But sometimes the truth hurts. There are truths that we might in some sense be better off not knowing.

A solution to the Puzzle of Promise makes two demands on our efforts: (1) that we articulate a conception of promise that explains how practitioners can acquire knowledge of it; and (2) that we articulate a species of epistemic warrant that elucidates how judgments of future promise can be justified, one which does not rely on the evidence of past problem solving and, indeed, which can be exercised "in defiance of" it, as Kuhn remarked. The preceding chapters satisfied the first of these demands. It is the remit of this chapter to address the second.

The principal "move" in my development of an alternative species of epistemic warrant is the transition to a new perspective on the nature of inductive practice. That transition involves drawing out the implicit classificatory acts that accompany any attempt to engage in inductive projection. Once this dimension of inductive practice is brought to the fore and is seen as a core element in any act of inductive projection, the emergence of our alternative species of epistemic warrant naturally follows.

I begin in section 7.1 by questioning the notion of epistemic warrant which conceives of justification as justification for belief. Building on the results of the last several chapters, and in particular on the reception studies of the physicist and historian Stephen Brush, I draw attention to some seriously deep challenges to the idea that the epistemology of science should focus on refining our understanding of how true beliefs about nature are acquired. Fealty to the notions of truth and reality is a natural and defensible response to the need to explain how science has been able to play the historical and social role that it does. I get it. I'm not sure Kuhn did, although I also can't imagine him *not* getting it, so who knows? My concern, one which Kuhn inspired and which is shared by an increasing number of philosophers of science, is that these notions have long been plagued by problems that make them unfit as centers of epistemological gravity in the context of science. As they crumble, losing more and more of their substance to well-intentioned efforts to retain them in some form, the affiliated institutions orbiting around them are flung out into increasingly dark epistemological territory. We need to seek alternative epistemologies of science that provide the same kind of explanatory stability for which truth and reality have been valued, without threatening to consume the epistemic virtues of science in a sociocultural void which treats scientific inquiry as one among many "ways of knowing" about the natural world. The practice of mathematics, which for much of its history developed in close relation to our best science, has for me been a highly instructive and fertile platform from which to launch the search for meaningful, epistemologically satisfying alternatives.

This search led me to look more closely at some of the implications of Henri Poincaré's observation that any act of classification is an act of analogy, implications that bear directly on inductive practice. From this discussion it will emerge that any inductive projection involves decisions about how features of experience ought to be classified. It then follows that at least *part* of what makes a projection inductively valid is that it is rooted in a valid classification. Now, if classification is fundamentally an act of analogy, then the validity of any classificatory act will depend on whether the analogy or metaphor that grounds it is warranted. But the conception of warrant associated with analogy is understood to be quite unlike that associated with inductive projection. In particular, the content of inductive warrant appears to be dominated by "the evidence from problem solving" in a way that analogical warrant appears not to be, because an analogy is warranted just in case it is *apt*. In section 7.4, I look at the content and behavior of judgments about aptness. It turns out that the central features of aptness judgments converge in a surprising way with the features of paradigm choice described in Chapter 2. This coincidence is not accidental. Viewed in this way, judgments regarding the aptness of a classification are to be regarded as epistemic, on account of their foundational justificatory role in inductive practice.

7.1. Evidence for What?

The problem I've set out to solve in this book is that of how researchers can be epistemically justified in their adoption of an approach to inquiry that has little if any previous success to recommend it. The reason this appears to be a problem is because of the presumption that the epistemic warrant for adopting an approach is a function of how successful that approach has been in the past. Even Kuhn himself implicitly appeals to this epistemological principle when he argues that, in the absence of a glorious past, the decision to adopt a new paradigm must be based on faith—that is, either the decision is based on past successes and epistemically warranted, or it's unwarranted.

This dogma is to the modern philosophical understanding of inquiry what metaphysically necessary foundations were to premodern epistemology of science. Even as perceptive a critic as Thomas Kuhn was unable to see his way out from underneath it. What we need is a way of thinking about epistemic warrant in the context of scientific inquiry that subordinates prior success such that "the evidence from problem solving" becomes one of many considerations that factor into the decision of which approach to adopt. Poincaré's epistemology of inquiry crystallizes the intuition that, fundamentally, what is at stake in the decision to adopt a new approach is how far a researcher believes he will be able to get by using it—how fruitful it will be. Sometimes past success will factor into this judgment, other times not. Sometimes (probable) truth will factor into this judgment, other times not. As we will see, Poincaré is often at pains to stress how little an approach's history or track record matters as a motivating factor in its continued use. This way of thinking about the psychology of inquiry cannot help but bring to the fore questions about what, precisely, evidence for or against a scientific theory is evidence for or against. By default, we are, I believe, accustomed to thinking of some evidence—some empirical fact—being "in favor of a theory" as being in favor of that theory's truth. If that is what is meant by "evidence for," then a lot of what governs decision-making in the context of theory choice is disqualified as evidence.

We can, however, imagine an alternative conception of "evidence for" which is considerably more catholic in the sources of evidence to which it appeals. If we think of "evidence in favor of" a scientific theory as evidence in support of its *use*, for example, then we are able to admit as sources of evidence a far greater range of considerations that are actually brought to bear on choices made in the context of inquiry—considerations which are "of importance in actual scientific decisions." That some framework better lends itself to visualizability than another might count very strongly in support of its use in inquiry, despite the fact that visualizability is orthogonal to truth. This same inclusive spirit applies to the other "epistemic virtues," especially those whose connection to truth is as tenuous as visualizability. I examine this issue in detail in Chapter 8. The point I wish to make here is that a wide range of considerations could in principle be brought to bear on the question of whether to use a scientific theory or broader approach to inquiry (paradigm, framework, etc.), one of which is (or at least *can* be) the probable truth of the theory.

Why are there more sources of evidence that are able to speak in favor of a theory's use? Plainly, because there is far more that is relevant to the advisability of a theory's use than there is to the probability of its truth. Pragmatic considerations about ease and efficiency, cognitive considerations about understanding and generality, pedagogical considerations about exemplification and transmissibility, aesthetic considerations about beauty and elegance, even sociological factors about prevailing disciplinary norms and cross-disciplinary affinities, and so on, and so on—all these and more can and frequently do simultaneously bear directly on a practitioner's decision to use/adopt a theory or approach, because each factor contributes in some way to the future development of inquiry. It is plausible to think that practitioners need, for example, a substantial track record of solved problems, a high degree of predictive accuracy, and a series of successful novel predictions to make a determination as to the probable truth of a theory. But it is not plausible to think that practitioners must possess each or indeed any of these elements before making an epistemically warranted decision regarding which theory should in the future guide research. All of the factors associated with the nonvertisitic categories listed above carry information regarding the extent to which a theory might promote the growth of knowledge. Appeals to these considerations are epistemic, because such appeals are used by practitioners to facilitate the growth of knowledge.

It is likely that these different kinds of considerations thrive in different epistemic environments. For example, perhaps pedagogical considerations of transmissibility are most advantageously influential at fairly advanced stages of theoretical development, when we might have several logically equivalent versions of a theory that differ only in their formal or aesthetic properties. Kuhn thought that aesthetic criteria were particularly influential during times of "extraordinary science," which are something like a combination of very advanced stages in the development of some theories and early embryonic stages in the development of others. It is, first and foremost, an empirical question which considerations tend to predominate at which stages, and there will doubtless be variation across practitioners with respect to what sorts of considerations they find most facilitative of further inquiry. We should expect both kinds of variation (across practitioners and across stages of inquiry), given that each stage (and for that matter, each problem) and each practitioner has peculiarities that call for different kinds of assistance.

Of course, there is no guarantee that such appeals will always result in the growth of knowledge. Allowing one's decision to be influenced by a preference for greater generality, for example, might backfire; it might result in a loss of accuracy or understanding, or it might produce nothing at all due to overgeneralization or trivialization. But then, neither is the appeal to the alleged indices of probable truth—track record, predictive success, and so forth—a surefire method for arriving at true theories. Indeed, given that we have rejected as false every theory (save some of our current ones) which could lay claim to a sterling track record and a sackful of predictive successes, we could reasonably assert that these properties are better indices of falsehood than they are of truth. Nevertheless, despite the distant glimmer of truth having repeatedly turned out to be a mirage, our pursuit and use of predictive theories, of theories that can boast of an impressive array of solved problems, and of theories that have directly resulted in the discovery of new phenomena, has consistently led to the growth of scientific knowledge. Conceived of in this way, what is traditionally taken to be evidence for a theory's truth might be more appropriately taken to be evidence for a theory's potential to promote the growth of knowledge.

This would make the alleged indices of truth a species of the genus to which pragmatic, aesthetic, cognitive, pedagogical, and sociological considerations belong—namely, the genus of factors that positively affect our ability to further advance our knowledge. It should come as no surprise, then, that this is exactly how Poincaré treats truth in the matter of hypotheses. Acknowledging in passing merely that a hypothesis "ought always, as soon as possible and as often as possible, be subjected to verification," he immediately points out that there isn't much value to a true hypothesis—that, indeed, a hypothesis that has failed an experimental test

has rendered more service than a true hypothesis. Not only has it been the occasion of the decisive experiment, but, without having made the hypothesis, the experiment would have been made by chance, so that nothing would have been derived from it. One would have seen nothing extraordinary; only one fact the more would have been catalogued without deducing from it the least consequence. (Poincaré 1946, 134)

Had the hypothesis been true—had it been repeatedly verified by experiment—we might never have discovered the extraordinary new phenomenon that was indicated by its actual experimental failure. It is only because the hypothesis has been experimentally rejected that we have been sent on the hunt for a hitherto undisclosed aspect of nature: "If the test does not support it, it is because there is something unexpected and extraordinary; and because *there is going to be something found that is unknown and new*" (Poincaré 1946, 134).

7.2. The Explanatory Failure
of Predictive Success

Poincaré's remarks about the virtues of false hypotheses over those that are true emphasize in some of the strongest possible terms the rather anemic contribution to the growth of knowledge that is made specifically by a hypothesis's truth. As he saw it, a false hypothesis carries with it the promise of future discovery. As a tool of inquiry, it will carry us a certain distance; but as long as we keep digging, something is bound to turn the spade, requiring us to develop some new way forward that will allow us to overcome the impasse.

By contrast, a true hypothesis offers no such guarantee. Even if we can be sure that it will carry us along the path of inquiry for a time, what lies at the end? We use the hypothesis to its fullest possible extent, and then—what? We move on to something else, just as we would had the hypothesis been experimentally rejected, only without the benefit of experimental failure to point us in a productive new direction. The mere notion that it cannot be experimentally refuted does not by itself portend a glorious future. Branches of inquiry do not need to be severed to die; they can also whither. Several philosophers have observed in this connection that what we want from scientific inquiry is not simply truth, but *significant* truth. But this cannot be quite right. For it seems that what we want is *significance* per se; whether the significance is attached to a truth or a falsehood is of little consequence—if anything, Poincaré notes, the falsehood is of greater service.

The dissociability of significance and truth is perhaps best illustrated by Stephen Brush's wide-ranging series of studies of the impact of novel predictive success on theory adoption (Brush 2015). If we accept for the sake of argument the scientific realist's view that novel predictive success is a proxy for truth (or approximate truth), the results of Brush's studies should give pause to anyone committed to the notion that probable truth is criterial—or even important—for theory adoption. There is a considerable amount of lore surrounding the notion that novel predictive success routinely makes a crucial difference to whether a theory is adopted. Historical data do not bear this out. Brush found that it really didn't matter one way or the other whether the theory had a record of successful predictions. In some cases, theories were adopted in the absence of predictive successes; in others, theories were ignored despite having been known to have made successful predictions. Sometimes, theories with novel predictive successes were adopted, but the stated reasons for adoption were unrelated to successful novel predictions. Instead, what mattered in each case was whether practitioners found that they could use the theory as (part of) a framework for conducting research.

This theme is illustrated with unique clarity in Brush's study of plasma scientist and Nobel laureate Hannes Alfvén. Alfvén made a number of successful novel predictions about the physics of the solar system based on "the postulate that plasma, consisting of electrically charged particles, is the most prevalent state of matter in the universe; hence most of the phenomena beyond the earth's atmosphere are to be explained in terms of plasma properties" (Brush 1992, 577). What is instructive about this episode is that, although the predicted phenomena are recognized by the scientific community, in none of the cases did the community take the prediction to have been confirmation of the theory that led to it. According to Brush's analysis, "the continuing resistance to Alfvén's work is based on a widely held opinion that his predictions are not derived from a plausible physical theory" (Brush 1992, 585). Because of this, as one of Brush's correspondents commented, they are not "anything a scientist can use in the development of understanding" (Brush 1992, 584). The framework from which Alfvén derived his successful novel predictions is so alien and unintuitive that the scientific community has no way to use it as a more generalized strategy for pushing inquiry forward.

Upon reflection, this pattern should have been the expected result. That is, we should expect a theory to appeal to practicing researchers only if they can envision how they might potentially use it to structure future inquiry. When they cannot see any means of doing that, they search elsewhere for a potentially productive framework. In other words, it might not matter to practitioners that a theory can boast of some successful novel predictions, *even if practitioners are convinced that novel predictive success can only be caused by true theories.* Practitioners are not rationally obligated to value predictive success just because truth alone can explain it. For the mere fact that a theory is true cannot guarantee that it will promote the growth of knowledge. It is not difficult to imagine a scenario in which some aspect of nature might be so complicated that a correct description of it would in principle be incomprehensible to the human mind. In such an instance, a true scientific theory would *ex hypothesi* be useless. Maybe practitioners would admire it simply for the fact that it is The True Theory, sort of in the way the apes venerate that obelisk at the beginning of *2001: A Space Odyssey*. But as far as the practice of science is concerned, that theory contains nothing of value. A perfectly plausible situation like this speaks, at least to me, decisively against the principle that what we seek from scientific inquiry is significant truth. With respect to this hypothetical aspect of nature, no truth could be more significant than The True Theory. And yet clearly its possession would be of no epistemic benefit to any practitioner. True theories—even significant true theories—are not an end in themselves. As the Brush studies show, they are at best treated as one among many potential means by which inquiry might be carried forward to a further stage of development.

Interestingly, in the broad array of reception studies he conducted, Brush could find only one instance in which novel predictive success was the decisive factor in motivating practitioners to adopt a theory: Mendeleev's periodic law. Why would it happen that the novel predictive success of the periodic law was epistemically significant but the novel predictive successes of, say, general relativity were not (Brush 2015, chapter 11)? I want to argue that, in the case of the periodic law, the specific form that its novel predictive successes took—literal gaps in the periodic table filled in by gallium, scandium, and germanium, along with the correction of some existing estimates of certain atomic weights—sent an unmistakably clear message to practitioners. Important properties of each of these elements had been foreseen by Mendeleev, and the confirmation of his predictions would have provided nearly irresistible motivation for practitioners to use the periodic law as a guide for generating further discoveries by filling in other gaps in the periodic table. The periodic law specified exactly where to look if one wanted to discover a new element (Brush 1996).

Thus, there is a good reason why the only instance of theory adoption driven by novel predictive success uncovered by Brush is the reception of Mendeleev's periodic law. None of the other episodes of theory reception he examined provided practitioners with as clear a signal of the potential bonanza of imminently solvable problems that awaited them as did the novel predictive successes generated by the periodic law. By showing that the periodic law could describe in advance what other elements there should be and what they should be like, Mendeleev enabled practitioners to treat each gap in the periodic table as a well-structured problem waiting to be solved, a problem whose solution would constitute a discovery of undeniably fundamental importance to our knowledge of the universe. As one might have predicted, novel predictive success mattered in this case because the specific nature of those successful predictions allowed them to be understood as referring to a whole family of solvable problems—namely, the remaining gaps in the periodic table. Those remaining gaps would have been seen as directly analogous to the earlier gap-filling exercises that generated the discoveries of gallium, scandium, and germanium.

We have been attempting to articulate an alternative conception of the epistemic significance of novel predictive success, because the standard philosophical view—viz., that novel predictive success is epistemically significant because it is caused by true theories—is not viable. Rather than deploy the standard response to this standard philosophical view—viz., that lots of false theories have made successful novel predictions—I have endeavored to sever the presumed connection between epistemic significance and truth. On the one hand, it is easy to imagine plausible scenarios in which practitioners might not see a theory's (probable) truth per se as a sufficient reason to adopt it, such as when a true theory is too complex to be useful. On the other hand, it is easy to imagine plausible research scenarios in which practitioners might have strong epistemic motivations to pursue a theory without regard to whether the theory is true or not. Fundamentally, researchers have an overriding interest in facilitating the growth of scientific knowledge, and this overriding interest is what governs appraisals of epistemic significance. Truth per se is neither a necessary nor sufficient condition for serving this interest.

Once one accepts that a theory can be either (a) true and epistemically insignificant, or (b) epistemically significant regardless of its truth status, a number of difficulties magically disappear. First, decisions about which paradigm should guide future research need not be presumed to be arational simply because they cannot be grounded in evidence of a theory's probable truth. It is possible for choices about what would facilitate the growth of knowledge to be epistemically rational and yet fail to track truth. Why? Because while truth may have a compulsory effect on our beliefs, it need not have such an effect on which tools we choose to employ in the context of inquiry. Philosophy of science has been dominated since its modern incarnation in the early twentieth century by the presumption that the most basic or essential form of a theory is fundamentally propositional and therefore truth-functional. I believe we now have sufficient evidence to conclude that this notion is misguided and is directly responsible for a great deal of confusion regarding how scientific knowledge actually develops. A theory is a cognitive tool that is amenable to a variety of representational modes. It often *can* be represented as a proposition; indeed, representing it as a proposition is a powerful strategy in certain contexts. But there are other contexts in which its most advantageous cognitive role is nonpropositional. This situation is analogous to the Cartesian (or Fermatian) representation of shapes as sets of mathematical points. There is obviously a lot that one can learn by representing a circle as $r^2 = (x_2 - x_1)^2 + (y_2 - y_1)^2$. But there is also a lot that one can learn by representing a circle as a specific geometric shape. And there is more to be learned by visualizing the circle as a conic section. Each of these nonpropositional forms of representation facilitates thinking about circles that may not be available—indeed, may even be prevented—by conceiving of them solely as loci.

Because propositional representation is just one form that a theory might take, the normative constraints imposed on belief by a theory's truth or falsity need not carry over to other cognitive states. There is no tension in believing a theory to be true and choosing not to use it, or in believing a theory to be false and choosing to use it. The rationality of use is subject to a different set of normative constraints than those governing the rationality of belief. Probably both sets play influential roles in inquiry.

Second, assuming that novel predictive success is a signal of a theory's truth, we can understand the messy historical relationship between successful novel predictions and theory adoption: maybe the historical fact that novel predictive success is neither necessary nor sufficient for theory adoption is a symptom of the fact that truth is neither necessary nor sufficient for theory adoption. In all the episodes examined by Brush save one, novel predictive success made no difference to theory adoption. This is compelling evidence that the epistemic value of novel predictive success is not contextually invariant. So, either:

(1) Novel predictive success doesn't track truth, or
(2) Truth is not a governing epistemic ideal in scientific inquiry

(or both). For philosophical reasons, I'm interested in exploring the consequences of (2), although I suspect that both (1) and (2) are correct. In the periodic law case, where novel predictive success *does* make a difference to theory adoption, its epistemic significance is better explained by the notion that exemplifying a family of solvable problems through which practitioners could expect to obtain concrete successes, rather than if we take them merely to reflect the theory's truth.

I've raised what I take to be the two most philosophically interesting problems associated with the standard view that novel predictive success is epistemically significant because it functions as a warrant for inferring a theory's (probable and/or approximate) truth. First, a theory's being true won't necessarily make it useful for scientific inquiry. So there must be more to the story of why practitioners value truth, when they do. Second, the idea that practitioners routinely and uniformly place value on novel predictive success is not borne out by the history of science. There is simply no way around that, and we need to all grow up and stop insisting that novel predictive success is epistemically significant whether practitioners realize it or not.

7.3. The Social Conception of Inductive Projection

Our journey up to this point has been guided by the general goal of articulating a species of epistemic well-foundedness that can reward decisions based upon it with the kind of epistemic payoff that is characteristic of scientific inquiry. The last several chapters have been aimed at developing a replacement for the canonical view that the epistemic payoff sought by researchers is the true belief and its attendant representational forms, propositions being chief among them. In place of that canonical view, I offered an image of science in which researchers are in search of solvable problems rather than

truths, and that those searches tend to terminate in the discovery
and cultivation of a more or less generalized way of structuring
problems to make them solvable. Such "frames," rather than true
beliefs about nature, are the epistemic payoff characteristic of scientific inquiry.

This alternative form of epistemic payment cannot help but lead
us to reflect on our conception of epistemic justification. Whereas
indices of truth, such as novel predictive success is purported to
be, are an appropriate basis for an epistemology of science centered around the idea that researchers are focused on justifying
decisions that are aimed at obtaining truths, those indices become
less appropriate once we abandon the notion that truths are the
payoff that researchers crave; in certain easily specified scenarios,
a fixation on indices of truth can impede the growth of knowledge.
I've related several episodes from the history of inquiry in which
researchers were rewarded for defiantly plowing ahead in the face
of an absence of such indices. They did not get lucky, nor was their
reward that of the "faithful" in Kuhn's de-rationalized epistemology
of science. They based their decisions on the promise of a fount of
well-structured problems, and those decisions paid off. They routinely do.

What we still don't have is a bespoke philosophical account of
epistemic justification that is tailored to the demands of this alternative form of epistemic payoff. The most fundamental of these
demands is that the novel form of justification be achievable by an
agent without the benefit of an established track record of solved
problems. We have by now seen lots of cases where a frame has
been adopted in the absence of a record of solved problems. What
we have not been given is a way of characterizing those choices that
clarifies a sense in which they are epistemically warranted.

The process of developing this alternative sense of epistemic
warrant begins with a careful reexamination of the notion of a
warranted generalization. In this section I argue that the species of
warrant involved in a warranted scientific generalization is that of

a *warranted analogy or metaphor*. This thesis, were it to be properly supported, has a number of advantages. First, it offers us a conception of warrant that can be achieved in the absence of a track record of solved problems. Second, it allows us to explain puzzling features of paradigm/frame choice in terms of a choice of metaphor or analogy, a perspective from which those features do not seem so puzzling. Third, it explains why decisions made on the basis of a warranted metaphor or analogy tend to result in a juicy epistemic payoff in the form of families of solvable problems. Lastly, it provides an alternative framework for reflecting on a much more general set of problems associated with inductive inference.

I want to begin our reexamination of epistemic warrant by reflecting on an interesting comment made by Poincaré in a chapter of *Science and Hypothesis* called "Hypotheses in Physics." Beginning with the observation that "We all know that there are good experiments and poor ones" (Poincaré 1946, 128), a modern reader might expect him to spell out this distinction in terms of rigorous experimental design, but Poincaré's epistemic sights are trained on more exciting vistas. In fact, the same experiment can be either good or poor depending on who is doing it, because what makes for a good experiment is not that it be designed and executed with all due diligence; a pupil's accurate reading of a thermometer ranks as a poor experiment, while a skilled physicist's reading of the same thermometer is a good one. For Poincaré, the latter is a good experiment because it "informs us of something besides an isolated fact; it is that which enables us to foresee, that is, that which enables us to generalize" (Poincaré 1946, 128). In other words, what makes an experiment good is not that the results be *reliable*; reliability is shared by good and poor alike. Rather, it is that the results be *generalizable*.

For Poincaré, the results of an experiment are generalizable if they can be extended to unexamined cases in a way that is generative of further insight. This is why the pupil's experimental run can be a poor experiment while the skilled physicist's run can be good, even if we assume that the experiment was performed in the exact same way in each case: only the skilled physicist has the store of broad-based background knowledge that would allow her to appreciate the significance of the results in a broader context. "From the first reading [the pupil's] we could not infer anything," he observes—not because the pupil's results are unreliable, or because they are at odds with other bits of our knowledge, but because the pupil is an ignorant schoolboy who lacks the cultivated sensitivity necessary for seeing the relationships between the new results and other knowledge (Poincaré 1946, 128).

On Poincaré's view, generalizability is a tripartite relation consisting of (1) the experimental results (the "facts"); (2) the formal relations between those results and other parts of our knowledge; and (3) the predilections and powers of particular human minds. As Goodman (1954) emphasized, each fact stands in myriad relations to other facts. The key to understanding generalizability— what he called the New Riddle of Induction—lies in solving the puzzle of why we are characteristically only interested in a very specific subset of those relations. I can't say whether Goodman's focus on entrenched predicates in particular would have resonated with Poincaré, but the conventionalist spirit of Goodman's solution would have undoubtedly pleased him. Both recognize the centrality of the practitioner's background knowledge and epistemic goals for explaining why the results of an experiment are projected in one direction rather than another.

Notice that, on this conception, an experiment need not *confirm* a generalization in order to be good; it only needs to possess features which practitioners can connect by analogy to other phenomena in a way that generates new lines of inquiry. Per our discussions in several chapters, this is because there is nothing that prevents a *disconfirmed* generalization, or a *failed* attempt to solve a research problem, from inspiring practitioners and from promoting within them a willingness to similarly structure other lines of inquiry, or future attempts to solve that same research problem. Knowing that a generalization is false as a description of a particular set of facts does not foreclose opportunities to use that generalization to guide the investigation of other problems, just as knowing that a generalization is true does not by itself *generate* opportunities to use that generalization to guide the investigation of other problems. As Poincaré remarks, "Whether verified or condemned, they [generalizations] will always be fruitful," so long as practitioners are *willing to use them* as a source for structuring other investigations (Poincaré 1946, 136). Echoing our discussion from the previous chapter, a good experiment for Poincaré is one which can serve as an exemplar: certain features of the experiment can be treated as referring to a class of related problems.

On the account I'm developing in this book, it is understandable why Poincaré invokes the property of fruitfulness in connection with practitioners' willingness to treat a generalization as a source upon which to model future inquiry. For the specific perspective described here toward experiments and their related generalizations is, as I argued previously, precisely the one involved in exemplification. A good experiment is one that enables practitioners to foresee, and whether they are enabled to foresee depends strictly on whether they are *willing* to extend aspects of the experimental result by analogy to other research problems. The unavoidable dependence on the practitioner for the generalizability of results is, for Poincaré, further compounded by the fact that the specific conditions that produce an experimental result are particular to a certain time and place, "circumstances . . . which . . . will never reproduce themselves all at once" (Poincaré 1946, 128). Therefore, the connection even between a specific experimental result and subsequent runs of the same experiment can only be made through analogy; individual runs bear enough similarities to one another that they can be classified as individual instances of the same phenomenon. While we might regard this as the kind of innocent and forgivable sort of analogy that is necessary for developing any kind of understanding of the natural world, its status *as an analogy* serves a crucial purpose in Poincaré's epistemology of science. For, by accepting that individual runs of an experiment can only be collected *analogically* under the same class, Poincaré thereby softens the epistemic blow of acts of classification/generalization based on seemingly more tenuous analogical connections. Fundamentally, all classification/generalization is made through analogy. Ergo, all warranted generalizations are made through analogy. So the project of distinguishing between warranted and unwarranted generalizations, between valid and invalid projections, becomes (in part) that of distinguishing between valid and invalid analogies.

As Poincaré would have known, the validity of an analogy is in the eye of the beholder(s), as it were, depending as it does on a balance between the degree to which it (1) gives salient features their due, (2) makes a compelling case for similarities between those features and features of the object of analogy, and (3) offers some useful insight. Thus, the validity of an analogy—and thereby a scientific generalization—is largely a matter of agreement within the community of practitioners: it is practitioners and their research traditions that determine which features are salient and whether the analogy treats those features appropriately; it is practitioners who must be persuaded of the case for relevant similarities between analogical subject and object; and it is practitioners who either will or will not draw insight from the analogy, and to whom that insight may or may not be useful.

A current debate in paleontology nicely illustrates Poincaré's radical alternative conception of inductive generalization.[1] I use this example because it was recently in the news, and because I've personally been involved in debates with practitioners, whose correspondence I relate here. However, this kind of example could have come from *any* discipline, because scientific disciplines are

---

[1] See, for example, https://www.cnn.com/2023/05/22/world/wildlife-crisis-biodivers ity-scn-climate-intl/index.html (last accessed June 1, 2023).

habitually confronted with the question of whether to continue employing a particular way of classifying phenomena.

A mass extinction is not just a really big extinction event. It is an extinction event with special evolutionary properties. Paleobiologists have sought to understand the specific sorts of extinction phenomena that are required to elicit those special evolutionary properties. Let us suppose, consonant with scientific practice, that the community of paleobiologist has settled upon a scientific characterization of the phenomenon they call *mass extinction*: for an event to be a mass extinction is for it to result in the extinction of 75 percent or more of globally widespread durably skeletonized marine invertebrate genera (GWDSMIG for short). This characterization, we like to say, "supports induction and explanation." That is, it licenses certain inferences about unexamined cases, such as that future (or current) extinction events that are not captured by this characterization will not involve the signature "change in rules" for evolutionary success that accompanies mass extinction events (Jablonski 2005). Or it allows us to make sense of, say, a period of unusually intense evolutionary innovation that occurred in the deep past. It is also specific and well-structured enough to do something that philosophers, to our enduring shame, have rarely thought to include among the epistemically important properties associated with the meanings of natural kind terms— namely, that they facilitate real scientific inquiry, projects that can be developed to a level of maturity worthy of the scrutiny of the relevant scientific community. We can use this characterization of mass extinctions to investigate, say, the potential causal relationship between the survival of globally widespread durably skeletonized marine invertebrate genera, on the one hand, and the reorientation of the rules of evolutionary success, on the other. Or we can try to understand, as David Jablonski has done, why properties like geographic range appear to shield some groups from the vortex of doom into which most species are cast during these tumultuous times (Jablonski 1986).

As with any good natural kind term, the virtue of the GWDSMIG characterization of *mass extinction* lies in the fact that (1) it captures the five truly epic extinction events reflected in the fossil record, as well as that (2) the simultaneous extinction of GWDSMIG has not occurred outside of these events. But it isn't flawless. It has in the past led to scientific disagreement over whether the current accelerated extinction rate is a true mass extinction—the *sixth* mass extinction (e.g., Barnosky et al. 2011). In 2012 at a workshop on extinction at Colgate University, I—partly in an effort to probe the limits and durability of the paleontological characterization of *mass extinction*, partly just to be annoying—asked the invertebrate paleobiologist Douglas Erwin whether it would qualify as a mass extinction if everything on Earth was dying *except* the GWDSMIG. His immediate response was, no, that would be a *biodiversity crisis*: "the term [*mass extinction*] is relatively well-defined in the paleontological literature. We (paleontologists) know quite well what a mass extinction is, thank you very much" (pers. comm., March 29, 2013). However, in a follow-up conversation, another workshop participant, the paleontologist Nan Arens, expressed the worry that

> The really interesting bit that I've not thought about before, is that the *definition* of mass extinction (that all us paleontologists know) was defined based on marine invertebrates with shells. Should we assume that it is the same for other lineages? We always have. Should we? The rules of evolution are sure different for plants, why not extinction too? If not, then we've got to put the modern diversity loss (catalogued mostly in things other than marine invertebrates with shells) through some other lens. (pers. comm., April 3, 2012)

The principal concern expressed here by Arens is that we might encounter an extinction event that cries out for classification as a mass extinction but which does not involve the simultaneous extinction of GWDSMIG.

Beneficiaries of Kripke's (1980) *Naming and Necessity* will recognize in this real-world case a reflection of the classic gambit in which a label and its canonically attributed properties come apart. These cases took the general form of the sort of scenario minted by Hilary Putnam in his classic paper, "It Ain't Necessarily So," in which

> Evolution has produced many things that come close to the cat but that it never actually produced the cat, and that the cat as we know it is and always was an artifact. Every movement of a cat, every twitch of a muscle, every meow, every flicker of an eyelid is thought out by a man in a control center on Mars and is then executed by the cat's body as the result of signals that emanate not from the cat's "brain" but from a highly miniaturized radio receiver. (Putnam 1962, 660)

Although Putnam's robot cats (or, say, Kripke's Gödel/Schmidt) are far-fetched, they capture an essential component of scientific practice with which we have been concerned throughout this book: both the members and the defining characteristics of natural kinds are—perennially—candidates for amendment. Our conception of a certain individual or a certain kind of thing centers around its association with some cluster of salient properties. When that association is put under some form of duress or another, we are forced to choose among (1) standing by the traditional conception and disqualifying certain ill-fitting cases on the basis of principle; (2) concluding that we were wrong about that individual's (or kind's) defining characteristics; or (3) concluding that there is, in fact, no such individual or kind, nor has there ever been. In the case of robot cats, Putnam's intuition—which I share and which generally coheres with the direction in which Kripke's cases were designed to lead—is that, in this scenario "we should continue to call these robots that we have mistaken for animals and that we have employed as house pets 'cats' but not 'animals' " (Putnam 1962, 660). In other words, we go with (2): we conclude that we were wrong about the defining characteristics of cats.

Now, not everyone responds to these cases in the same way; Katz (1975, 96), for example, says we should conclude that cats don't exist. To me, this kind of disagreement shows that the examples are doing what they're supposed to be doing. In the context of scientific practice, we should expect there to be disagreement within the community over how to resolve these kinds of conflicts—that is, conflicts in which we seem to be pulled simultaneously toward two incompatible options (violations of Kuhn's "no-overlap" principle). Determining whether we are in the midst of the sixth mass extinction is not simply a matter of comparing current extinction patterns with the definition of *mass extinction*. It is, as Goodman (1954) observed, more a matter of balancing the inferences that we're intuitively disposed to accept with the principles of inference that we're loathe to reject. And while our principles of inference do—*in principle*—license certain inferences about what members of the natural kind *mass extinction* can be expected to be like, those principles will be subject to amendment if they impede the inclusion of certain extinction events about which we feel particularly strong. Kripke-Putnam intuitions notwithstanding, there is no way to predict how these conflicts will get resolved in the context of scientific inquiry. They will be sorted out at the community level, in response to the prevailing demands of inquiry (or through other, even more erratically behaved social mechanisms). Members of the community will engage in reasoned debate in conversation, at conferences, and in the pages of journal articles. Eventually, they will reach consensus. The debate over the meaning of *mass extinction* is not unique in this respect. Communities of both natural scientists and mathematicians routinely revise the boundaries of their categories in response to confrontations between existing boundaries and compelling reasons to violate them.

This is a data point with which any serious account of the way in which natural kinds "support induction" must reckon. Inductive projections are traditionally conceptualized as attributions of certain properties to a certain unexamined object. The epistemic warrant for those property attributions comes from our assignment of the object to a class each of whose members has the properties in question. For instance, my inference that the next emerald I observe will be green derives its warrant (it is supposed) from my acceptance of the generalization, "All emeralds are green," which is itself ultimately grounded in the empirical fact that all observed emeralds have been green. However, viewed from a perspective that acknowledges the pervasive and substantive role of classificatory choices, an inductive projection is not simply an attribution of certain properties to a certain unexamined object. It is fundamentally a statement regarding a research community's future propensity to privilege those properties over other considerations when classifying the unexamined object. For example, suppose we choose to frame our empirical generalization not as one about emeralds per se but rather about the history of their classification, as in, "All observed emeralds *have been classified as* green." This alternative version achieves the important result of emphasizing the fact that all empirical generalizations are, at root, descriptions of a history of sustained classificatory practice, a history which the standard formulation—"All observed emeralds have been green"—misleadingly suppresses.

In point of fact, the standard formulation goes beyond our evidence in a way that the classification-based version humbly refuses to do. For our evidence for the claim that all observed emeralds have been *classified as* green is more direct than is our evidence for the claim that all observed emeralds have been green—perhaps some of them have been classified as green without having actually been green. The epistemic warrant for asserting "All observed emeralds have been green" is the historical fact that all objects classified as emeralds have been classified as green. If things had been otherwise—say, if some objects that were classified as emeralds were also classified as blue *despite having actually been green*—then we would have no warrant for asserting that all observed emeralds have been green, *even though they, in fact, might all have been green* (we just didn't know it, and for whatever reason, classified some as blue). On the other hand, had some observed emeralds (unbeknownst to us) *actually been blue*, our warrant for asserting "All observed emeralds have been green" would not change so long as it remained true that all objects classified as emeralds have been classified as green. Why? Because our warrant for assertion rests on that history of classification, and nothing requires that history to reflect any mind-independent truths about the actual properties of actual emeralds. Thus, either "All observed emeralds have been green" means, in the context of scientific practice, "All objects classified as emeralds have been classified as green," or we do not know that all observed emeralds have been green.

Accordingly, when I inductively project that the next emerald I see will be green on the basis of the fact that all emeralds are green, the content of that projection need not be construed as an attribution of the property of greenness to unexamined emeralds. It could equally—and, I think, more accurately and faithfully—be construed as an inductive inference of the following form: "All objects that have been classified as emeralds have been classified as green. Therefore, the next object to be classified as an emerald will be classified as green." In other words, rather than being conceptualized as projections about the mind-independent physical world, *inductive projections are more appropriately understood as projections regarding how likely it is that the research community will come to some form of agreement regarding* whether some object ought to be classified as both green and an emerald. One of the principal strengths of the Putnam-Kripke examples is that they drive a wedge between an object and its putatively necessary properties in a way that makes the research community's role in the inductive process easier to discern. For, if the question of whether being an animal is necessary for being a cat turns partly on the research community's response to the shocking revelations concerning the innards of what we've been calling "cats," then the warrant for inferring that the next cat we examine will be an animal cannot simply rest on what we think cat-hood requires in some cosmic sense. Apparently, it must also rest on our estimation of the community's future propensity to classify something as a cat given that it has been classified as a robot (so, not an animal). Our estimation of that propensity will be informed partly by the community's prior classificatory practice, but not entirely; that is why there is a question as to how the "robot cats" case might be resolved.

Upon what else does that estimation depend? In general, the other ingredients that go into shaping that propensity vary from discipline to discipline, and they are highly sensitive to social, historical, and investigative context. The debate over the (alleged) sixth mass extinction nicely illustrates this. Despite the fact that all previous events classified as mass extinctions were also classified as having involved the simultaneous extinction of GWDSMIG, it is an open question whether the next event classified as a mass extinction will be classified as also having involved the simultaneous extinction of GWDSMIG. There is intense pressure from both within the discipline and from broader social institutions to classify the current extinction event as a mass extinction. We do not know what, if anything, we will lose in terms of predictive and explanatory power by modifying the current boundaries of the concept *mass extinction* in a way that decentralizes durably skeletonized marine invertebrates. And we do not know how the community will choose to proceed when confronted with a particular loss of explanatory or predictive power. We don't know, for example, what they would choose if confronted with the choice between (a) categorizing the current extinction event as a mass extinction and (b) maintaining their ability to explain why the determinants of evolutionary success change following a mass extinction.

In the previous chapter, we developed the Poincaré-inspired thesis that any act of classification is an exercise in analogy-making: to assign some object membership in a certain class is to adopt the position that the object's degree of resemblance to other class members is sufficient to warrant inclusion in their special club. We are now in a position to begin unraveling the epistemic significance of this perspective. When the community of paleontologists is deciding whether to classify the current extinction event as a mass extinction, they are assessing whether the current extinction event bears the right sort of analogy to previous mass extinction events to itself be classified as a mass extinction. Part of that assessment involves the current event's similarities with previous mass extinctions. But the assessment draws on considerably more than that. It also involves a reevaluation or a recalibration of what sorts of similarities *matter* for making that analogy work. Previous mass extinctions have all involved GWDSMIG. But how much do we care about GWDSMIG, *really*? Would it be such a big deal if we suspended the notion that an analogy to mass extinctions strong enough for membership in that class is only warranted if the subject of the analogy involves the extinction of GWDSMIG? Maybe it would; maybe it wouldn't. Whether that sort of analogy *ends up being* deemed as warranted will depend on the outcome of the research community's reassessment of what sorts of similarities matter to them. This easily detectable feature of scientific experience is, I believe, the primary reason why Kuhn could not bring himself to embrace the "context of discovery"/"context of justification" distinction. Communities of practicing scientists simply do not abide by this distinction, however compelling it may be philosophically. Nor can they. As the "robot cat" cases illustrate so well, we cannot know how committed we are to the centrality of some property as a necessary or defining characteristic of a given class until we encounter situations in which our commitment to that necessity is put to the test. No amount of a priori reflection can determine how those tests will turn out. Although I share Kripke's and Putnam's intuitions, not everyone does. And scientific inquiry is *a lot* more complicated than these toy examples. The history of science makes it clear that no one should trust their intuitions regarding how such conflicts will be resolved.

The picture that is emerging is one in which an inductive projection is a projection not about the properties of an unexamined object, but about a research community's decision to adopt a certain sort of analogy between members of an established class and some object not (yet) included in that class. Our earlier discussion indicated that whether the research community will accept the validity of, say, the analogy between past mass extinctions and the current extinction event is not as straightforward as making sure the current extinction event has the properties required for a mass extinction; what counts as "required for a mass extinction" depends on a host of other factors, such as the salient properties of the current extinction event, as well as the research priorities of the community at a given moment. When members of the community inductively infer that the next mass extinction will involve the simultaneous extinction of GWDSMIG, they are betting that the analogy between previous mass extinctions and the next mass extinction will be deemed valid if and only if the latter involves the simultaneous extinction of GWDSMIG.

The question is, what would motivate them to make such a bet? In other words, why not bet that future iterations of the community will decide to characterize an event as a mass extinction despite the fact that GWDSMIG continue to thrive? One is tempted here to give a traditional sort of answer of the form, "They make the bet because they believe that the inductive generalization is true—that is, because they believe that mass extinctions necessarily involve the simultaneous extinction of GWDSMIG." But this traditional sort of answer is inadequate, because the truth of that generalization only guarantees the stability of the association between the simultaneous extinction of GWDSMIG and *being* a mass extinction. While the truth of that generalization might be able to ensure the stability of the correlation between those two properties, it cannot guarantee the *stability of the community's practice of* caring about that correlation—that is, of the community's practice of drawing an analogy between previous mass extinctions and other extinction events on the basis of the simultaneous extinction of GWDSMIG.

So what *would* guarantee the stability of that community practice, such that its members could, with justification, expect that practice to be preserved across future iterations of the community? This book's obsession has been developing and substantiating a general answer to that question—namely, that a research community's confidence in the long-term stability of any community practice comes from the perception of its *promise*, that is, from their confidence that that practice will continue to bear fruit. A true generalization promises to remain true, but it does not ensure that practitioners will continue to have something to work on. While the *truth* of the generalization "Mass extinctions involve the simultaneous extinction of GWDSMIG" might depend on the stability of the correlation between the simultaneous extinction of GWDSMIG and being a mass extinction, the *continued fruitfulness* of analogizing previous mass extinctions to other extinction events on the basis of GWDSMIG most certainly does not. The continued fruitfulness of that analogy depends on whether it will continue to facilitate scientific inquiry by generating families of solvable problems.

To conclude: an inductive inference is an inference concerning the perceived longevity of a certain approach to classification, one which says to classify something as an *F* if it is sufficiently analogous to other members of the class of *F*s. The longevity of such an analogy is rooted in its fruitfulness—that is, in its propensity to generate families of solvable problems. As long as it continues to be fruitful of solvable problems, it will tend to persist. The justification for an inductive inference thus turns out to lie in whatever properties of an analogy make members of the research community confident that that analogy will continue to be fruitful.

Look, I'm as unhappy about the epistemological messiness of science as anyone. It would be much easier to piece together a philosophically compelling account of scientific knowledge if we could just build it from first principles and sense data and whatever, like we used to do. Such an account, however, would be about something other than science, if it was about anything at all. For better or for worse, scientific inquiry is conducted by real people. People make inferences not on the basis of what the world is like—whether emeralds *really are* green—but on the basis of what they *judge* the world to be like. Practitioners of mathematics and the natural sciences add a further inconvenience to an already unhelpful epistemological morass, in that they must solve problems. They do not have the luxury of proceeding solely on the basis of what they judge the world to be like. From among the permissible ways of representing the physical and mathematical world, they must privilege approaches to representation which possess a peculiar propensity to facilitate the production of solved problems.

7.4. The "Aptness" Alternative

In this section, I argue that judgments regarding the fruitfulness potential of an analogy are rooted in judgments of *aptness*: in the context of research, to assert that an analogy or metaphor is apt is to claim that it has the potential to facilitate research by generating families of solvable problems. I first make some general observations about the general relationship between aptness and warranted classifications. I then look at the behavior of aptness judgments. I also attempt to articulate the specific content that aptness judgments express in the context of scientific research. In section 7.5, I highlight the alignment between features of aptness judgments and features of paradigm choice. There I defend the idea that the alignment exists because paradigm choices are made on the basis of aptness judgments. This alignment, in turn, reveals the alternative sense in which decisions regarding which framework should in the future guide research are epistemically justified: the choice of paradigm is justified to the degree that it provides an apt framework for research.

What makes a characterization "apt"? Consider the exchange at the beginning of the last chapter from the sitcom *Parks and Rec*. Leslie's response references two kinds of criteria required for aptness: a descriptive one (One *could* say that) and a normative one (But *should* one?). Part of what makes a characterization apt is its descriptive accuracy. The philosopher Walter Ott responded to my email congratulating him on his fascinating study, *Causation and Laws of Nature in Early Modern Science*, saying. "Thanks. It took hours." One *could* say that. But it's not exactly an apt characterization of the time he invested in writing it. I occasionally will introduce my wife in the following way: "This is Maysan, my first wife." I *can* say that, because she is. *Should* I? No. No, I should not. Paul Dirac is reported to have once greeted a visitor to his home by saying, "Allow me to present Wigner's sister, who is now my wife." He's not wrong; that woman definitely was Eugene Wigner's sister. This story survives because it is a comic *mischaracterization*. But notice that even the concept of a *mis*characterization implies accuracy of description. After all, one can mischaracterize a set of facts. Mischaracterizing a set of facts does not involve stating inaccuracies. Rather, it involves characterizing facts in a way that fails to satisfy certain additional criteria by which we judge the aptness of a characterization.

What are these additional criteria? Or, at least, what kind of criteria are we asking for? The mischaracterizations listed here suggest one very broad theme: each of them seems to violate certain expectations or norms of assertion that are supplied by context. In particular, they appear to be norms governing the assignment of salience, or degrees of salience. Ott's technically accurate description of how long his book took to write is comical for the way in which it violates our default expectations for the size of the units in which book-writing time should be expressed. It violates an ostensible norm of casual conversation that says that a quantity should be expressed in whichever units allow for the smallest magnitude greater than one (that's probably not exactly right, but it's close). It is like saying, "Saturn is inches away." Book-writing intervals and planetary distances are associated (at least in our language) with certain appropriately sized units. Introducing Maysan as "my first wife" violates our expectations for the conditions under which one should mention the order of appearance in which one's wife happens to fall (viz., when one has been married more than once and is no longer married to any earlier wives). In my situation, Maysan's status as my first wife is decidedly not salient. (It may even be so nonsalient that its nonsalience is itself salient!). When Dirac leads the introduction of his wife with a reference to Wigner's sister, he violates our expectations for the priority ordering of her properties in that context, an ordering which would have been preserved if Wigner had uttered that statement instead of Dirac ("Allow me to introduce my sister, who is now Dirac's wife."). In that conversational context, her relation to Dirac is expected to be treated as more salient than who her brothers and sisters are.

In each case, there seem to be properties that context fixes as particularly salient. Mischaracterization occurs when that degree of salience is not given its due. Importantly, the salience of these properties is itself dependent on a rich background of firmly established community norms that dictate how information should be characterized, norms which could easily have been otherwise. Ott's joke would not have landed had the *hour* been our only unit of time; it just would have been a bizarrely irrelevant statement (because, of course, it took hours. What else *could* it take?). My unique brand of spousal humor would get even fewer laughs were we to live in a society in which divorce and remarriage every five years was compulsory. Dirac's statement would have been a perfectly sensible human utterance had Eugene Wigner been, say, Supreme

Dictator of Earth ("Wow! Wigner's sister. Such an honor! Did you say she was also your wife?").

But not every context possesses salience assignments that are so firmly fixed, clearly grasped, or widely shared. One reading of the *Parks and Rec* exchange reflects such a context. When Leslie Knope asks whether one *should* "say that a book is nothing more than a painting of words, which are the notes on the tapestry of the greatest film ever sculpted," she can be interpreted as drawing attention to the fact that it is not clear whether such a characterization would be consistent with the norms governing salience in that context. Or she could be expressing genuine uncertainty as to what the salience norms are for that context, thus making it difficult to discern whether they would be violated by such a characterization. Alternatively, she might be interpreted as drawing attention to the fact that such a characterization clearly *does* violate the salience norms for that context (making the question rhetorical). Even under this interpretation, however, the NPR host's asking whether the characterization is apt reflects that this context's salience norms are not entrenched enough to make a clear and rapid determination of aptness. To ask whether a characterization is apt is, in part, to ask whether it is consistent with contextually relevant salience norms; or whether its consistency with contextually relevant salience norms can be discerned.

A picture of aptness is beginning to emerge according to which two kinds of criteria seem to dominate. Part of what makes a characterization apt is its accuracy. Another part of what makes a characterization apt is that it pays due deference to the operative norms for salience in a given context. But the "painting of words" example adds a wrinkle to this picture, in that it illustrates the way in which a sufficiently novel characterization can expose gaps in, destabilize, or potentially *influence* the system of norms that are supposed to be governing it.

The Maxwell episode in Chapter 5 provides a concrete example of the kind of circumstances in which such a phenomenon might arise. According to the analysis I developed there, the right way to understand Maxwell's complaint about the state of electrical science being unfavorable to speculation is to view it as a claim about the absence of a frame for "arranging and interpreting the results of previous investigations." Such a frame would perform this critical epistemic function by crystallizing a certain perspective on what kinds of features ought to be treated as salient in the context of electrical research, allowing researchers to develop characterizations of electrical phenomena according to these norms of salience. Because electrical science at that time lacked a serviceable research frame, a practitioner had no obvious rationale for viewing a specific cluster of physical and quantitative features of electrical phenomena as particularly salient. Thus, he had no means for distinguishing between a characterization that was apt and one that was not. In sum, what bothered Maxwell about the state of electrical science was that there was no agreement on what an apt characterization of electrical phenomena would even look like.

Maxwell's advocacy of the fluid frame was an explicit proposal to treat as salient (1) certain dimensions of electrical force (direction and intensity) and (2) certain features of electricity's behavior that were analogous to the behavior of fluids. To accept his proposal would have been to adopt a set of salience norms for governing the characterization of electrical phenomena. But adopting that set of salience norms would not thereby confer aptness upon a characterization that was consistent with them. Maxwell is not simply advocating for the fluid frame out of deference to his eighteenth-century predecessors. Rather, he is advocating it *because* he thinks it provides for apt characterizations of electrical phenomena. The norms of salience are what make it capable of generating characterizations; that's just what a frame *is*. Something more is required to make those characterizations *apt*.

This "something more" was the set of critical epistemic functions that Maxwell saw electrical science as having no resources to perform in its present state. Not simply arranging and interpreting results—that much is supplied by the salience norms—but *doing so in a way that facilitated further inquiry*. Thus, his advocacy of a frame is not merely advocacy of a set of salience norms, but of the promise that those salience norms will enable practitioners to perform critical epistemic functions. What made fluid-based characterizations apt was that it would "bring before the mind, *in a convenient and manageable form*, those mathematical ideas which are necessary to the study of the phenomena of electricity."[2]

Similar lessons can be derived from the species-as-particles episode. The triumph of the species-as-particles frame is not encapsulated simply in its providing a new perspective on the salience of spindle-shaped diversity graphs or on the salience of clumps in morphospace. Although it *did* prescribe new salience norms (by, for example, reorganizing the conceptual relationship between clumpiness and natural selection), the motivation for adopting these new norms centered on their capacity to facilitate the investigation of causal hypotheses in paleontology through a particular characterization of a zero-force evolutionary state. As with Maxwell's advocacy of the fluid frame, practitioners understood the facilitative capacity of this approach to the fossil record well in advance of any tangible results or solved problems. They could determine the aptness of these zero-force state characterizations in advance of any concrete successes because they could see both their accuracy and their ability to "bring before the mind, in a convenient and manageable form," the means by which causal hypotheses could be *investigated*—structured in such a way as to make problems solvable.

We have arrived at a picture of aptness according to which what makes a characterization apt is that it facilitates further inquiry by employing norms for constraining and structuring research problems, norms which put those problems into a "convenient and manageable form." Before moving on, I want to pause to draw

---

[2] Maxwell (1890, 157; my emphasis).

attention to certain features of this picture which bear on the main epistemological problem which has motivated this entire study— that is, how theory choice can be rational in a way that explains the success of science but without ultimate regard to a theory's truth. Practitioners know that much of their success in inquiry depends on something as simple as being able to structure problems in a convenient and manageable form. Convenience and manageability are properties that can be judged more or less instantaneously and, in particular, without regard to any past history of concrete success; the motivations governing Maxwell's adoption of the fluid frame are the same as those behind so many choices regarding certain styles of notation and diagrammatic representation. When practitioners detect convenience and manageability in a problem structure, they can know immediately that a very steep hill on the epistemic landscape has been summited, because they now have an understanding of how to structure problems in a way that will make them solvable. Convenience and manageability belong to a family of related notions which share a number of properties that are causally efficacious in the development of scientific and mathematical knowledge. Principal among these properties is the immediacy with which they can be detected. I explore this family of properties in further detail in the next chapter.

7.5. Apt Metaphors and the Progress of Science

In his seminal paper on metaphor in science, Richard Boyd drew a profound contrast between the behavior of metaphors in literature and their behavior in science. As he observed,

> a literary metaphor has its "home," so to speak, in a specific work of a specific author; when the same metaphor is employed by other authors, a reference to the original employment is often

implicit. When the same metaphor is employed often, by a variety of authors, and in a variety of minor variations, it becomes either trite or hackneyed, or it becomes "frozen" into a figure of speech or a new literal expression. . . . Literary interaction metaphors seem to lose their insightfulness through overuse: the invitation to explore the various analogies and similarities between the primary literal subject and the metaphorical secondary subject becomes pointless or trite if repeated too often. (Boyd 1979, 488)

The behavior of a metaphor in contemporary literature is highly proprietary. It is almost antithetical to the function of a metaphor in that context that it should achieve widespread usage.
In stark contrast to this phenomenon lies the behavior of metaphors in science:

scientific metaphors, on the other hand, become, when they are successful, the property of the entire scientific community, and variations on them are explored by hundreds of scientific authors without their interactive quality being lost. They are really conceits rather than metaphors—and conceits which extend not through one literary work, but through the work of a generation or more of scientists. (Boyd 1979, 488–489)

We have seen numerous examples of this phenomenon already, so I won't belabor it here. But Boyd's emphasis on the infectiousness of scientific metaphors as compared with literary metaphors is a novel data point. What is it about successful scientific metaphors that makes them so irrepressibly invasive?

I believe there are several reasons for this. Part of the explanation lies in a major difference between the sciences and the contemporary literary arts[3]—namely, that within the sciences there is

---

[3] *Pace* Boyd, Curtius (1953) shows that literary metaphors were treated as common property for most of literary history. See discussion in Haufe (2023, chapter 7).

an intense demand for tools that will aid in problem solving. As a tool for generating apt characterizations, science provides a highly favorable context for the spread of a given metaphor, because there is a level of demand for apt characterizations in science that has no parallel in the literary arts. But this cannot be the complete story, because the intense demand for representational vehicles in the sciences cannot by itself explain why *particular* metaphors become widespread, rather than that there be a vast panoply of metaphors that are idiosyncratic to particular researchers. In other words, the motive and opportunity to use metaphors could just as easily result in each practitioner using a different metaphor. But this is not what we see. Instead, we see entire communities converging on the use of a small set of representational vehicles, often with surprising speed.

The key to this puzzle lies in the nature of aptness judgments. Aptness judgments exhibit an astonishing degree of uniformity. If you think a metaphor is apt, it is likely that other members of your linguistic community will as well. This phenomenon is so well-established that it is simply presupposed in much of the experimental work on metaphor, so much so that the aptness of a metaphor is basically treated as an objective property; it is not just a matter of taste. Linguists, for example, will compare the speed with which subjects derive the meanings of familiar apt metaphors versus those of unfamiliar apt metaphors—not by asking subjects to rate aptness, but simply by constructing apt metaphors beforehand. Researchers are confident that if they consider a metaphor to be apt, respondents will, too. Such is the objective purport of aptness.

As Boyd observes, the successful use of literary metaphor in fact *depends* on the ability of writers to reliably presuppose that readers will judge their metaphors to be apt. The whole system is built on the presumption of uniformity of aptness judgments. This presumption reflects an understanding that there is a large extent to which a metaphor's aptness is an objective property in the same way a sentence's grammaticality is. While there is some threshold level of competence required to detect grammaticality per se, people who possess that competence exercise it as effectively with novel sentences as they do with sentences they've heard before. Similarly, subjects characteristically converge on the aptness of novel metaphors, and they process novel, unfamiliar metaphors as quickly as they do familiar metaphors when the metaphors are apt (Blasko and Connine 1993).

Let us suppose that this uniformity in aptness judgments also holds in the context of science and mathematics, as indeed appears to be the case. It would then follow that, when a metaphor is particularly apt, it is likely to be seen as particularly apt by the community of practitioners at large. Now, I argued earlier that for a characterization to be apt in the context of rational inquiry means that it facilitates further inquiry by employing norms for constraining and structuring research problems which put those problems into a "convenient and manageable form." If (1) that is what an apt characterization amounts to in science, and (2) aptness judgments tend to be uniform across community members, then it should follow that (3) community members will tend to converge in the judgments concerning a characterization's ability to facilitate further inquiry by putting problems into a "convenient and manageable form." Like literary aptness, "convenience and manageability" is something that can be judged more or less immediately, without preparation, and without a prior history of success or familiarization.

This would explain why certain metaphors and other framing devices tend to quickly become widespread and persist across generations of practitioners. If a representational vehicle is particularly suitable for facilitating further inquiry by putting problems into a "convenient and manageable form," it will be quickly perceived as being so by the community of practitioners. Given the intense demand for representational vehicles that are effective in this way, particularly apt metaphors (or similes, or analogies) should be expected to rapidly make their way into widespread use.

I believe that this simple and direct explanation is of supreme importance for understanding the rationality of choices made in the context of scientific inquiry. First: if a frame's aptness is something that can be judged reliably and quickly even when it is unfamiliar (as in metaphor), then we can see how the distinction between (a) a frame that will be particularly effective in facilitating inquiry, and (b) a frame that will be less effective can be made quickly and reliable by practitioners with no knowledge of or concern for whether the theory embodied by the frame is true, or probable, or has a track record of problem-solving successes. The rationality of judging a frame's ability to facilitate inquiry should be understood along the same lines that we use to assess the rationality of aptness judgments outside of science. What sort of evidence do speakers require to determine whether a literary metaphor is apt? They draw on their knowledge of the vernacular in which the metaphor is expressed. That is all. It is not as if they need to check to see if the metaphor has been effective in the past to make a reliable determination as to whether it is apt. Firmly established results in linguistics make it clear that aptness can be judged immediately by competent speakers. We should expect this competence to exhibit itself with respect to the more specialized vernacular of research communities. I will have more to say in detail in the next chapter regarding what sorts of properties of frames positively influence aptness judgments in science.

Second: we can see how grounding one's choice of frame on the basis of aptness can make an enormous difference to one's ability to eventually solve problems. To say that a frame is apt is to say that it generates apt characterizations. An apt characterization of a problem is a particular way of structuring the problem so as to make it solvable. If a practitioner judges a frame to be apt—which she can do quickly and reliably—she has determined that it structures problems in a way that makes them solvable. In this way, the selection of a particularly apt frame eventually produces the concrete successes for which scientists are normally rewarded.

7.6. Conclusion

The aptness of a frame—a metaphor or an analogy—can be confidently assessed without reference to history. When a frame is judged to be apt in the context of research, the content of that aptness judgment specifically relates to the inherent potential of a given metaphor or analogy to generate families of research problems that are well-structured enough to make them solvable. For a researcher to judge a frame to be apt is for her to deem it prone to providing apt characterizations of phenomena and, thus, able to provide things for her to work on.

There now needs to be some account of the sense in which aptness judgments qualify as a species of epistemic justification. I have followed Poincaré and Goodman in conceiving of scientific and mathematical generalizations as analogies. It is a corollary of this commitment that a warranted generalization be conceived of in terms of some kind of "warranted" or "valid" analogy. We are now in a position to flesh this conception of warrant in more detail.

The use of a frame is warranted when that frame is judged by a practitioner to be apt. When a frame is judged to be apt, it is judged (1) to be accurate; (2) to give salient properties their due; and (3) to generate apt characterizations—that is, to structure problems in a way that is convenient and manageable, and that facilitates further inquiry. Because aptness judgments in research contexts involve an assessment of a proposed problem structure, those judgments are keyed to the production of families of solvable problems. And because aptness judgments are based on properties that can be perceived directly and immediately—through the same kind of competence that supports the capacity to detect apt metaphors generally—those judgments can be made reliably. On this view, then, there is a species of epistemic warrant in science and mathematics which consists ultimately in a practitioner's exercise of his professionally developed sense of what sorts of frames tend to best facilitate inquiry.

What *kind* of warrant *is* this? The traditional sense of epistemic warrant appeals to a warrant for *belief*. That is clearly not a good fit for understanding the kind of epistemic warrant associated with an apt analogy or metaphor—as reflected, for instance, in Maxwell's efforts to discourage the interpretation that he was in some way attempting to support the assertion that electricity was a fluid. Rather, the kind of warrant attached to aptness judgments involves a warrant for *use*. The use of the fluid frame is warranted in the sense that it is (1) accurate; (2) it gives salient properties their due; and (3) it generates apt characterizations. This alternative sense of warrant does not appear to give rational grounds for belief in any straightforward way—at least, not for belief in the literal truth of the frame. "The Queen Mary entering Walden Pond" is J. D. Salinger's apt metaphor for the tinnitus-inducing vocal stylings of Ethel Merman; its use is warranted. That warrant does not give me grounds for believing that Ethel Merman might, in reality, be the Queen Mary entering Walden Pond.

But there are reasons to think that the traditional sense of epistemic warrant—at least the one that is relevant to rational inquiry—was never about belief in the first place. Goodman's argument in "The New Riddle of Induction," for example, teaches us that the generalization, "All emeralds are green" possesses more warrant than the generalization, "All emeralds are grue" because *green* is an entrenched predicate whereas *grue* is not. Now, there is no more evidence for the hypothesis that all emeralds are green than there is for the hypothesis that they are grue. That is what makes Goodman's Riddle interesting. However, the firmly entrenched place that *green* occupies in our language makes for a more apt characterization of emerald color than does the alien predicate *grue* . While both are *accurate*, *grue*violates the conventional norms of salience governing color attribution, and it does a poor job of facilitating further inquiry; in no way does thinking of emerald color as grue lend convenience and manageability to our quest for understanding.[4]

---

[4] Thanks to Marc Lange for discussion on this point.

Science is not merely awash in metaphor. Scientific descriptions of the natural world are *inescapably* metaphorical. Accepting that these descriptions are metaphorical does not require that we relinquish the idea that they are epistemically well-founded. There is a well-understood sense of well-foundedness associated with metaphor—*aptness*. It explains how choices made on its basis tend to produce the payoffs familiar from the history of science. Its estimation is native to the practice of inductive projection, properly conceived as a projection regarding a research community's approach to classification. And its detection reflects much about actual scientific decision-making that makes no sense from a traditional epistemological perspective. I examine these unusual decision-making properties in the next chapter.

8

Aptness and the Causal Role
of Epistemic Virtues

*Imagine what would happen in the sciences if consistency
ceased to be a primary value.*

—Thomas Kuhn, "Postscript," 186

*Change the list [of values], for example by adding so-
cial utility as a criterion, and some particular choices will
be different, more like those one expects from an engineer.
Substract accuracy of fit to nature from the list, and the enterprise that results may not resemble science at all, but perhaps philosophy instead.*

—Thomas Kuhn, "Objectivity, Value Judgment,
and Theory Choice," 331

Even in his most generous mood, Kuhn never deigned to explore
the causal roles of the various epistemic virtues whose general relevance he would eventually come to accept. These are literally the
only statements I've ever found which, subject to a suitably char-
itable interpretation, suggests that it occurred to him that there's
something special about the specific set of values that modern science has adopted.[1] In case there was any lingering doubt, in neither case does the text give any indication of what he thought might

---

[1] Thanks to Marc Lange for pointing me to the second passage.

Fruitfulness. Chris Haufe, Oxford University Press. © Oxford University Press 2024.
DOI: 10.1093/oso/9780197666395.003.0008

happen were the values of consistency or accuracy to be abandoned in the sciences. The sentence before for the first quote, for instance, makes the observation that "the developmental pattern of the plastic arts changed drastically when representation was abandoned as "a primary value"; the sentence after moves on to another point entirely. That there is *some* causal role is an idea he deems acceptable, but there was never any suggestion that those causal roles explain anything about the unique power of modern science.

At the beginning of this book, I noted the widely observed causal influence of aesthetic properties at crucial nodes in the process of inquiry. For reasons related to the preservation of science's cognitive authority, there is a premium placed on being able to explain why aesthetic properties enjoy influence over the direction of inquiry in a way that highlights their ability to promote the satisfaction of paradigmatically epistemic goals. One explanatory tradition has attempted to do this by attributing to aesthetic properties a causal role in the production of true beliefs.[2] Owing principally to the mysteriousness of an alleged connection between truth and aesthetic properties such as visualizability, abstraction, elegance, symmetry, and the rest of the aesthetic lot, efforts to epistemically explain the causal influence of aesthetic preference in science have been unsatisfactory. What I hope to do here is provide the foundation for convincing alternative epistemic model of the role of aesthetic preferences in science. Building on Quine's early remarks about fruitfulness, along with close attention to scientific debates that turned explicitly on aesthetic issues, I argue that aesthetic criteria preferred by practitioners in mathematics and the natural sciences are partly constitutive of a frame's aptness, and that the epistemic contribution of aesthetic judgment lies in its ability to promote fruitfulness by increasing the solubility of problems. In contrast to the mystical beauty-truth connection, I emphasize the connection between aesthetic appeal and cognitive

---

[2] See Schindler (2018) for a recent systematic effort.

effort—Maxwell's "convenience and manageability"—to show why aesthetic preferences could be expected to play an important epistemic role in rational inquiry.

Additionally, we now have seen several instances in the history of science in which a theoretical framework appears to have been embraced by the scientific community well in advance of the accumulation of a stock of evidence that could be described by traditional metrics as rationally compelling. The ostensible lack of an epistemic inducement of sufficient magnitude leading to these transitions has naturally resulted in claims about the potential causal influence of nonepistemic factors, claims which, if substantiated, would present a serious challenge to the cognitive authority of science. Our task here will be, again, to explain how such instances can be explained epistemically, without appealing to the satisfaction of epistemic criteria that are traditionally associated with well-founded inference. I argue that, because of the connection between aptness and fruitfulness, the symptoms of potential fruitfulness are detectable at very early stages in the lifespan of a framework of inquiry, be that framework methodological or theoretical. One important reason for this is that fruitfulness potential has very much to do with what we can reliably project about our own cognitive capacities and the conditions under which we are most effective at solving scientific problems; this is precisely what aptness judgments track.

The general arc of this chapter's argument proceeds from the platitude that tasks often get easier as we get more familiar with them, and it ends with the stunning revelation that the aesthetic properties of theories/frames facilitate the structuring—and, thus, the solving—of research problems. The keystone of that arc is the observation that the ability to understand a metaphor is affected by that metaphor's aptness: an unfamiliar but apt metaphor is processed with the same speed as a familiar metaphor. One's facility with the use of a research frame is made possible by features that contribute to that frame's aptness, even when one lacks the experience required to become familiar with it. Researchers can make epistemically warranted choices about frames without prior experience, because epistemic virtues are constituents of the aptness of a frame. These virtues achieve in an instant what can otherwise only be achieved by increasing one's familiarity and experience with a research frame.

8.1. From Familiarity to Understanding

I want to begin arguing for the idea that aesthetic preferences are causally related to fruitfulness by first elaborating the connection between familiarity and fruitfulness to which Quine (1955) briefly alluded in his "Posits and Reality." According to Quine, familiarity's role in the promotion of fruitfulness is mediated by its facilitation of what he calls "effective thinking." I think this is roughly right, but it is also very roughly articulated. I would like to preserve this insight from Quine—in suitably modified form—because of its salutary emphasis on the idea that a theory's fruitfulness is most closely associated with our cognition, rather than with the theory's truth. To make better use of Quine's insight, I first describe the manner in which familiarity facilitates effective thinking, and then connect effective thinking to the particular conception of fruitfulness we've adopted—viz., the propensity to generate families of solvable problems.

Let's start off by considering the more general notion of *effective use* and then move on from there to derive specific lessons for the notion of effective thinking. The connection between familiarity and effective use is well-known to us in everyday life, as, for example, with regard to tools. Using a tool for the first time can often be a cognitively exhausting experience, not to mention a painful one. Anyone fortunate enough to have enjoyed the bliss of using power tools is acquainted with the "learning curve" that accompanies them. Even if the manner in which they are to be used is self-evident, one cannot just pull it out of the box and start wielding it like a boss. In my own experience, even something as simple as a hammer and nail, or a drill, will usually require a lot of concentration to use properly and effectively if I haven't used it in a while. For example, if someone asked me right now to cut six inches off a two-by-four with a radial saw, I could probably do it, but I wouldn't be able to talk to anyone while I was doing it. But after using the tool consistently for a certain length of time, we can typically execute complex tasks without devoting our full cognitive effort to them.[3] After a few hours, the same job can be done with a fraction of the initial cognitive effort, and often with better results.

The impact of familiarity on effective use is also well-known to experimental scientists, and its significance for the epistemology of science has been of peripheral philosophical interest science the seminal work of Michael Polanyi in his (1958) *Personal Knowledge*. One notable development of Polanyi's insight was the argument made by Harry Collins (1985) that replication of experimental results is difficult or impossible because of the fundamental role that familiarity with an experimental system plays in using it to achieve specific ends. As with the tools we use around the house, familiarity with an experimental system constitutes what Polanyi called *tacit knowledge*—things we know but can't articulate. Tacit knowledge of a variety of aspects of the experimental system's behavior is an essential ingredient in the effective use of the system, such that those who lack it have little chance of successfully executing an experiment, let alone achieving the degree of success enjoyed by more experienced users.[4]

The centrality of familiarity to effective use of an experimental system was one of the more prominent patterns observed by

---

[3] Professional builders can usually knock a nail in straight and flush with two strikes of a hammer. Compare the average Habitat for Humanity volunteer at sixteen to thirty strikes!

[4] For a convincing example, see Collins (2001).

Holmes (2004) in his study of investigative pathways. There he confirmed Hans-Jorg Rheinberger's evocative claim that

> [t]he elaboration of the capabilities of an experimental system 'requires familiarity with the system. This process takes time, which helps to explain why experimenters, once they have established their experimental work[,] often stick to it in almost symbiotic fashion.' (Holmes 2004: 123; quoting Rheinberger 1997)

Indeed, familiarity is valued so highly for its contribution to effective use that it is common to find more senior scientists using technologies which those in the embryonic stages of their scientific career consider wildly out of date. While some instances of the "symbiosis" described by Rheinberger are mere stubbornness, the nature of the case is probably rather one where the scientist estimates that the degree of experimental success per unit of cognitive effort she enjoys with her familiar system is higher than it would be were she to switch to more up-to-date technology. And, compared with her probable degree of success during the learning period for the new technology, she is most likely correct. Using her old system, she can effortlessly achieve with precision what would be overly demanding were she to make the switch, much in the way Frege described the "lignification" of ideas. Closely related to this is the phenomenon wherein scientists that are trained early on with the use of a new technology wield it far more effectively than their mentors.[5]

With familiarity's causal connection to effective use firmly in place, we can now go on to look more closely at its specific influence on the effectiveness of thinking. The classic means of exploiting this influence is through the use of analogy, where we attempt to make it easier to understand one thing by depicting it as a version

---

[5] See, for example, Sepkoski (2012) on the comparative skill of mentors and graduate students in paleontology with the use of computing technology.

of something with which we're already well acquainted. Darwin famously used this rhetorical mechanism to great effect in the *Origin*, where he draws on readers' familiarity with the manner by which new domesticated varieties are produced to help them understand how new varieties and species arise in nature. What had previously been, in Lyell's terms, the "mystery of mysteries" is instantly transformed into a relatively innocuous and relatable extension of a causal process already well understood by uneducated laymen.

Using analogy as a model for the connection between familiarity and effective thinking elegantly illustrates the status of effective thinking as a special case of effective use. In the latter case, we saw how the result of familiarity was to reduce the cognitive effort required to achieve a specific practical end. Thus, we should expect the impact of familiarity on effective thinking to be one which reduces the cognitive effort required to achieve some distinctively cognitive end. Now, if we (1) accept analogy as an appropriate exemplar of how familiarity promotes effective thinking, and (2) accept that analogy achieves its effect by using familiarity to make things easier to understand, then our expectation is confirmed: at least in the case of analogy, familiarity reduces the cognitive effort required to achieve the distinctively cognitive end we call *understanding*.

The notion that we promote our understanding of a given phenomenon by increasing its familiarity is fairly transparent. Independently of what I've argued earlier, it is natural to conceive of the process by which understanding is achieved as a limit process characterized by successive gains in familiarity—much as I attempted earlier to depict expert competence as a limit process characterized by successive gains in familiarity with a given tool or experimental system. Our initial state, when some phenomenon is totally foreign to us, is one in which the maximum amount of cognitive effort is required to achieve the sort of intuitive grasp associated with understanding. As familiarity increases, the amount of cognitive effort required to achieve understanding falls. In the limit, at maximal familiarity, understanding is achieved with minimum cognitive effort.

8.2. The Epistemology of the Solvable Problem

To summarize the argument up to now: true to Quine's observation, familiarity facilitates effective thinking by reducing the amount of cognitive effort required to understand a given phenomenon. We now want to see how to get from this result to the production of families of solvable problems. The concept of a *solvable* problem refines the traditional focus on solved problems in that it allows us to retain the explanatory power of the broader focus on "problem solving" while providing a critical link to the sort of future-directedness that must necessarily play a dominant role in scientists' decisions concerning how inquiry is to be organized. Scientific achievement develops in stages, beginning with the attempt to identify a genuine problem and ending with its solution. As the example of Darwin's analogy between breeding and adaptation nicely illustrates, an important node in the developmental process of scientific achievement lies in providing a successfully identified problem with some kind of "structure," a structure that promises to eventually lead to the problem's solution. The way in which an approach structures a problem is used as evidence of the approach's "future promise," because the pace at which we can solve a problem is strongly dependent on whether and how we structure it. This fundamental fact about scientific inquiry is well captured by an observation made by Herbert Simon, who probably thought more about problem solving than anyone before or since: "there is merit to the claim that much problem solving effort is directed at structuring problems, and only a fraction of it at solving problems once they are structured" (Simon 1973, 187).

In the remainder of this chapter, I want to work toward establishing my main thesis that the influence of aesthetic preference on the direction of scientific inquiry is *epistemic*, and that the epistemic contribution of aesthetic properties in science lies in the powerful, perhaps unique, role they play in a frame's ability to generate apt characterizations—that is, to provide structure to whole classes of problems. If science is epistemically special because of its distinctive ability to solve problems, and if science's ability to solve problems depends largely on its ability to structure them, contributions to problem structuring have to count as epistemic contributions. This shows that the aesthetic judgments play an epistemic role in science. And because problem structuring is fundamentally about solubility, the contribution of aesthetic properties lies specifically in the production of solvable problems. In this way, we obtain an epistemic explanation for the influence of aesthetic judgment in science, an epistemic explanation that rests on the contribution of aesthetic properties to fruitfulness.

The argument for this thesis proceeds through four steps:

1. Continuing with the line of argument which attempted to draw a richer, tighter, and more precise connection between familiarity and fruitfulness, I now try to show that, as a matter of historical fact, familiarity has been important for generating families of solvable problems.

2. I then argue more specifically that the reason why familiarity has been so effective at generating solvable problems is because it allows us to *structure* problems in ways that reduce the cognitive effort required to solve problems.

3. Following this, I describe cases where scientists have used the structuring capabilities of familiarity *prospectively*, as a way of justifying choices about what form a theory ought to take. This is significant for our purposes, because it provides a clear

illustration of how theory choice can be epistemically rational *even when* the choice demonstrably does not rest on which theory appears more probable, given the available evidence. Rather, these rational theory choices rest on perceptions of fruitfulness potential informed on the basis of structural properties of the theories.

4. Finally, I argue that certain aesthetic properties are pursued for the same fundamental reason as familiarity, which is a kind of aesthetic property, in a way. It is significant that only a rather limited subset of aesthetic properties has been found to influence decisions in the context of rational inquiry. One does not often hear, for example, of a theory's vivacity being a source of attraction for scientists. Rather, the small set of preferred aesthetic properties are all closely tied to structuring problems in some way, and the way in which each is tied to structure straightforwardly highlights that property's epistemic significance. Given the enormous impact of these aesthetic properties on our ability to structure problems, we should be surprised *not* to find them playing a powerful role in scientific inquiry. A useful contrast here is with the property of novelty or originality, which is typically viewed as a type of aesthetic property. If familiarity is valued because of its problem structuring abilities, it would be reasonable to expect novelty per se—on account of its inability to contribute to problem structuring—to be quite ill-favored. And indeed it is. The reason novelty per se is not used in the detection of fruitfulness potential is because, by its very nature, it cannot give scientists information about what future science will look like if we adopt whatever novel idea or method is being proposed. Novelty's contrast with familiarity and its structuring abilities refines this earlier point to give us a more precise sense of why novelty is uninformative for gauging fruitfulness potential.

8.2.1. Familiarity and Solvability

We can further bolster the connection between solved problems and scientific understanding by noting the similar role played by familiarity in each. Beyond suggesting that the convergence between scientific understanding and the solved problem is not accidental, this brings us closer to our current goal of showing how familiarity facilitates the more substantive version of fruitfulness I've been arguing for. The causal relevance of familiarity to problem solving is straightforward, and I certainly don't want to pretend to have discovered something important about this connection that is not already widely acknowledged. What I hope to do instead is highlight how knowledge of this connection is exploited within scientific inquiry for the purpose of reducing the amount of cognitive effort required to eventually solve problems. Familiarity causes fruitfulness by reducing this requisite cognitive effort and thus contributing to the propensity of problems to be solved—that is, by increasing their solubility.

The first step in illustrating this relation was taken already in Chapter 6, where I quoted a passage from Poincaré in which he asserts that solving a mathematical problem is of no value "if it has not enabled me to foresee the results of other analogous calculations, and to direct them with certainty, avoiding the blind groping with which I had to be contented the first time." Poincaré is specifically concerned with mathematical practice here, but I take this to be a important insight into the nature of progress in inquiry more generally. Expressed in these terms, the unit of scientific achievement is not the solved problem per se. Rather, it lies somewhere between the solved problem and the unsolved problem. For Poincaré, a solved problem is only of value if it highlights relevant similarities between itself and "a much more extensive class of problems." These similarities are what allow us to "foresee the results" of successful solutions to that more extensive class of problems, "avoiding the blind groping with which [we] had to be contented the first time." We know that this more extensive class of problems is solvable—and we know what form the solutions will eventually take—because of the problems in this more extensive class all possess an analogous structure to the one we just solved. Our solved problem is a sample of that class.

Readers acquainted with Kuhn's "Second Thoughts on Paradigms" will recognize its instructive overlap with Poincaré's view, for he makes a related point regarding the value of problem sets in textbooks. Physics students, he observes, typically find it difficult to take the formalisms they learn in textbooks and immediately apply them to the problems at the end of a given chapter, no matter how firmly they may have grasped the logic of the formalism itself:

> Almost invariably their difficulty is in setting up the appropriate equations, in relating the words and examples given in the text to the particular problems they are asked to solve. Ordinarily, also, those difficulties dissolve in the same way. The student discovers a way to see his problem as like a problem he has already encountered. Once that likeness or analogy has been seen, only manipulative difficulties remain. (Kuhn 1977, 305)

According to Kuhn, the value—and the intended purpose—of problem sets is to enable the student to identify subsequent problems as versions of a problem he already knows how to solve.

Both Poincaré and Kuhn have explicitly invoked the importance of the scientist being able to see an analogy between different classes of problems, and their respective points regarding the value of these analogies echo our earlier discussion of Darwin's analogy between the process by which new domestic varieties are cultivated and that by which new species originate. In the same way that Lyell's "mystery of mysteries" comes to be seen through Darwin's analogy as simply another version of the familiar breeding problems that we already know how to solve, Poincaré's "more extensive class of problems" and Kuhn's daunting problem sets are similarly converted through analogies with now-familiar problems into more or less manipulative exercises that require comparatively little cognitive effort. This corresponds to what is sometimes referred to as "plug and chug" work when dealing with equations. Problems can be seen as having a certain structure, and once an individual's ability to identify that structure has sufficiently developed, the only real challenges involve keeping the details straight in a given instance. What these examples indicate is the particular significance of achievements that bring the scientist to a point where "plugging and chugging"—or some more sophisticated version of it—is possible or appears to be on the horizon. Such events are ones in which whole groups of problems are suddenly rendered solvable because of their newly achieved familiarity.

Poincaré's assertion that "I shall have gained nothing for my trouble" if a solved problem fails to render other problems solvable is probably extreme, but it does bring out nicely the relatively immense value of generating solvable problems as compared with merely solving them. The significance of this point is plainly illustrated by Kuhn's "problem sets" example, since the value of conquering a particular problem for the student's progress pales in comparison to that of being able to conquer entire sets of hitherto unencountered problems. But this transition is also significant in the context of full-blown scientific inquiry, because the privilege of treating mysterious, complex, or confoundingly difficult problems as mere instances of a type of problem that we're already adept at solving is no less valuable for the working scientist.

Newton's approach to the full range of physical problems provides a compelling illustration, since it highlights several dimensions in which familiarity contributes directly to solvability. What Newton gives us is geometric proof after geometric proof, neatly organized around diagrams containing all the familiar lines, points, and shapes that we associate with geometry. These were the tools that composed what I. Bernard Cohen called "the Newtonian style," whereby Newton would depict a physical object (e.g., a planet) as an idealized mathematical entity (e.g., a sphere) and then subject that mathematical entity to familiar (and sometimes not so familiar) mathematical techniques (Cohen 1983, chapter 1).

While not to deny the tremendous power that the *Principia* displayed in, *inter alia*, its successful unification of the motions of terrestrial objects with those of the heavens, it is difficult to imagine the same degree of success being achieved without the assistance of the geometric framework that Newton chose to adopt. The reason for this is that, unlike the problems of motion in general, problems in geometry were ones that his contemporaries knew how to solve; more importantly, they knew when a result had been successfully proved. When Newton (like Galileo before him) framed physical problems in the purely mathematical terms familiar from geometry, he brought with him the full power and stature of geometry's problem-solving apparatus (Westfall 1980). These principally included a canonical form of diagrammatic representation and a universally recognized standard of validity (geometric proof). Not only did the Newtonian style allow researchers to see their way through the problems of motion as never before, but it told them what form successful solutions could be expected to take (Grosholz 2007). This latter element is critical for appreciating the appeal of Newton's approach in advance of most of the confirmed explanatory successes that were to come. We seek to understand the motion of objects. Geometry tells us the general conditions under which we can claim to have achieved understanding. If we can treat the problems of motion as a species of geometry problem, we will instantly grasp what it would be like to solve them. Most significantly, we can make sense of the influence the *Principia*'s Book II, which reported solutions to problems of fluid dynamics that, in contrast to Book I's results in mechanics, were not generally accepted by practitioners (Truesdell 1968, chapter 3).

The facilitation of solvability by familiarity is equally central to inquiry in the biological sciences, as several well-known instances from recent biology make clear. Much of Robert MacArthur's pioneering work in ecology employed this strategy. To take what is probably the best known example, MacArthur, along with E. O. Wilson, suggested the idea of treating the number of species on an island as an equilibrium point determined by the balance between rates of immigration and extinction (MacArthur and Wilson 1967). Much of the appeal of this highly influential approach to ecology was located in the fact that the problem-solving framework for equilibrium processes was by that time well established. So long as a plausible argument could be made for the role of equilibrium constraints in these biological cases, an approach which furnished ecologists with a robust, time-tested problem-solving toolkit was not a hard sell. By contrast, understanding the ecology of islands by looking at the specific biological details of its inhabitants is a project which still faces enormous challenges today.

Many more examples could be adduced, but only to marginally enhance what should be a sufficiently clear general point: familiarity is routinely used in science to generate classes of solvable problems. I take this phenomenon to be a species of the same genus of phenomena canvassed earlier, in which familiarity with a given task facilitates the pursuit of certain ends by reducing the amount of cognitive effort required to achieve those ends. In science, the focal end is the solvable problem—a problem whose solution is achievable through currently available means, or means whose availability appears inevitable.

8.2.2. The Prospective Use of Familiarity

I have been stressing the apparent fact that scientists use their established ability to solve certain kinds of problems to reduce the cognitive effort required to solve classes of unsolved problems. Based on the evidence that can be marshaled in support of this tendency, I believe it is plausible to consider the primary value of a solved problem to lie, as Poincaré says, in whatever ability it might have to illuminate potentially effective approaches to whole classes of problems. I have previously argued that this is equivalent to, in my terminology, generating families of solvable problems.

If the primary value of a solved problem lies in what it allows us to do with other classes of problems, we might expect there to be debates specifically over which of two (or more) acknowledged solutions to a given problem best facilitates the production of families of solvable problems. Given what I have argued earlier concerning the connection among familiarity, cognitive effort, and solvability, we might further expect that such debates will often hinge on the relative familiarity of each of the acknowledged solutions. Solutions with more familiar structures will tend to be preferred, because they reduce the cognitive effort required to solve a given problem and thus promote solvability.

We can see this prediction borne out in the recently published transcripts of the famous 1927 Solvay conference, where the world's most prominent physicists gathered to address what were at that time the deepest problems in quantum mechanics. One such problem was the fundamental indeterminacy of the behavior of particles posited by Neils Bohr. Einstein's fierce opposition to this feature of the theory is well known. But the great Hendrik Lorentz also opposed it, for reasons which were, like Einstein's, unrelated to the empirical success or mathematical coherence of the theory. At the conference, Lorentz vividly described his resistance to an interpretation of quantum mechanics which prevented him on grounds of principle, as Bohr's Copenhagen interpretation did, from forming a mental "image . . . absolutely sharp and definite" of the behavior of electrons:

We wish to make a representation of the phenomena, to form an image of them in our minds. Until now, we have always wanted to form these images by means of the ordinary notions of time and space. These notions are perhaps innate; in any case, they have developed from our personal experience, by our daily observations. For me, these notions are clear and I confess that I should be unable to imagine physics without these notions. The image that I wish to form of phenomena must be absolutely sharp and definite, and it seems to me that we can form such an image only in the framework of space and time. (Bacciagaluppi and Valentini 2009, 432)

These sentiments are useful for the clarity and variety of the preferences expressed, for their indication that Einstein's dissent was not just another one of his endearing peccadilloes, and especially for the easily discernible importance that Lorentz placed on certain kinds of familiarity for the practice of science. The "ordinary notions of time and space" are intimately familiar to us—"perhaps innate"—and are the basis of our ability to form mental images of the phenomena. This ability, at least for Lorentz, is essential for the practice of science. Without ordinary notions of time and space, mental representation of the phenomena is not possible. And without mental representation of the phenomena, physical science is *literally* unimaginable.

The importance of visualizability for the future of physics would also come up in a discussion of the relative merits of Schrödinger's wave mechanics over those of Heisenberg's matrix mechanics, two mathematically equivalent approaches to the description of quantum states. Schrödinger argued at Solvay that his newly minted wave equation possessed an important virtue over Heisenberg's matrix mechanics, in that it offered more "hope of achieving a three-dimensional conception" of the behavior of electrons:

What Mr Heisenberg has said is mathematically unexceptionable, but the point in question is that of the physical interpretation. This is indispensable for the further development of the theory. Now, this development is necessary. For one must agree that all current ways of formulating the results of the new quantum mechanics only correspond to the classical mechanics of actions at a distance. . . . If I am not satisfied with the current state of the problem, it is because I do not understand yet the physical meaning of its solution. (Bacciagaluppi and Valentini 2009, 428)

For Schrödinger's part, dissatisfaction with the cognitive obstacles raised by *a particular way of thinking* about quantum mechanics was enough to register fundamental disagreement with that approach, even though each acknowledged that it was, in his words, "mathematically unexceptionable." Rather, what was exceptionable about matrix mechanics was that it did not readily lend itself to representation in three dimensions, thus making it difficult to work with as a physical model. Being as they were two different ways of saying the same thing, the empirical fates of the two formalisms were inextricable: any data that conflicted with one would necessarily conflict with the other. But in Schrödinger's view, the fate of physical science looked very different depending on which of the two would serve as the basis for future inquiry. In this he was not alone. Lorentz diagnosed the problem as "merely a question of knowing which of the two representations is the most suitable, which is the most convenient" (Bacciagaluppi and Valentini 2009, 443)—and he had given the palm to Schrödinger in that regard: "if I had to choose between your wave mechanics and the matrix mechanics, I would give preference to the former, owing to its greater visualizability."[6] Heisenberg himself had recently commented directly on the importance to quantum physics of a formalism's susceptibility to three-dimensional representation, observing that the present quantum theory possessed "the disadvantage of not being directly amenable to a geometrically visualizable interpretation, since the motion of electrons cannot be described in terms of the familiar concepts of space and time."[7]

---

[6] May 27, 1926, Lorentz to Schrödinger; quoted in Miller (1987).
[7] Born et al. (1926, 558); quoted in Miller (1987).

The reason why determinism and three-dimensionality were viewed as "indispensable for the further development of the theory," I would suggest, was because they were "most convenient"; given their intimate familiarity with the deterministic, three-dimensional conception of physics on which they'd been raised, physicists believed they could wield it much more effectively than a conception which required them to abandon these familiar notions. It was thus believed to lend itself most easily to quantum theory's further development. For the group of physicists involved in the developmental stages of quantum theory, being able to use "the familiar concepts of space and time" to organize the investigation of quantum states meant being able to potentially convert problems in quantum mechanics into relatives of the problems in classical mechanics that they knew well how to solve. Gazing out into the open future of physics, they justifiably felt that the best chance of making progress on the new problems raised by quantum mechanics would be to stay broadly within the problem-solving framework that they'd been refining over the past three hundred years.

Poincaré too shared the view that future progress in physics would be facilitated through the use of a set of tools which "enables us to obtain a clear comprehension of the whole," for this is "what causes a fact to give a large return" (Poincaré 1910; quoted in Gray 2013, 18). Just as the early quantum theorists had done, Poincaré set great store by the power of familiarity to continue generating solutions to problems in physics, even despite the emergence of empirically superior alternatives. Indeed, so convinced was Poincaré of the central importance of the "return" on our investment in an idea that he even went so far as to predict (correctly, as it turns out) the enduring role of Newtonian mechanics in physics, just as the relativistic picture—notably, a picture to which he himself made an early and important contribution—was coming into view:

Perhaps, too, we shall have to construct an entirely new mechanics that we only succeed in catching a glimpse of, where, inertia increasing with the velocity, the velocity of light would become an impassable limit. The ordinary mechanics, more simple, would remain a first approximation, since it would be true for velocities not too great, so that the old dynamics would still be found under the new. We should not have to regret having believed in the principles, and even, since velocities too great for the old formulas would always be only exceptional, the surest way in practice would be still to act as if we continued to believe in them. To determine to exclude them altogether would be to deprive oneself of a precious weapon. I hasten to say in conclusion that we are not yet there, and as yet nothing proves that the principles will not come forth from out of the fray, victorious and intact. (Poincaré 1910; quoted in Gray 2013, 106–107)

In a related discussion, commenting on the rising influence of non-Euclidean conceptions of space, Poincaré lays bare the exclusive significance for inquiry of the "return" provided by our theoretical commitments:

What shall be our position in view of these new conceptions? Shall we be obliged to modify our conclusions? Certainly not; we had adopted a convention [that is, that space is Euclidean] and we had said nothing could constrain us to abandon it. Today some physicists want to adopt a new convention. It is not that they are constrained to do so; they consider this new convention more convenient; that is all. And those who are not of this opinion can legitimately retain the old one in order not to disturb their old [familiar!] habits. I believe, just between us, that this is what they shall do for a long time to come. (Quoted in Gray 2013, 112)

There is no need, he suggests, to abandon the conclusions derived from our commitment to Euclidean spatial geometry, because that commitment, like Minkowski's non-Euclidean version, is held conventionally. Our commitment to each is governed not by its truth, or by its compatibility with other commitments, but rather by its "convenience." For him, conflicts between conventions occur not at the level of meaning, but at the level of expected return on our investment in them. True to Quine's observation, these expectations were fueled by Poincaré's perceptions of simplicity and familiarity. As he observed early on (1907):

> It quite seems, indeed, that it would be possible to translate our physics into the language of geometry of four dimensions. Attempting such a translation would be giving oneself a great deal of trouble for little profit, and I will content myself with mentioning Hertz's mechanics, in which something of the kind may be seen. Yet it seems that the translation would always be less simple than the text, and that it would never lose the appearance of a translation, for the language of three dimensions seems the best suited to the description of our world, even though that description may be made, in case of necessity, in another idiom. (Reprinted in Poincaré 1914, 113; my emphasis)

Incidentally, Poincaré remained faithful to the Euclidean framework to the last, a fact which Poincaré scholar Scott Walter attributes—quite apart from Poincaré's general predilection for fruitfulness—to his perception of its ability to increase "scientific productivity" (Walter 2009, 215).

Thus far we have been concerned with establishing two points:

1. Familiarity facilitates the production of solvable problems by reducing the cognitive effort required to solve a given problem.
2. The power of familiarity to facilitate problem solving is actively exploited by scientists and mathematicians charged with the decision of how to approach a given family of problems.

A principal virtue of these facts is that they allow us to place into a broader positive epistemic framework the observation reported in Chapter 2 that novelty does not positively impact assessments of potential fruitfulness. It is easy to see why, if scientists are sensitive to the fact that familiarity is a "precious weapon," novelty would be resisted unless the old familiar approaches simply could not be made to work tolerably well with respect to some class of problems. Indeed, it is difficult to imagine conditions of inquiry under which that might fail to obtain. Relatedly, we do not need to appeal to stubbornness or cognitive bias to explain resistance to changes in theory and in technology. Rather, we can use the progressive effects of familiarity to show how such resistance is the default expectation from the perspective of epistemic warrant.

8.3. Solvability, Structure, and Aesthetics

I have been focusing on the use of familiarity in science because it is easily detectable and easily understood from an epistemic perspective. I now want to step back from the specific focus on familiarity in an attempt to catalogue other features of the design of scientific inquiry that enjoy similarly widespread appeal, for similar reasons. I've argued that familiarity holds its special place in inquiry in virtue of its ability to reduce the amount of cognitive effort required to solve a given problem. If that's correct, then other aspects of scientific practice can be expected to be equally popular, insofar as they have similar effects on cognitive effort.

Among the most storied and most mysterious aspects of scientific practice is the influence of aesthetic considerations on theory choice. Familiarity is recognized as one such aesthetic consideration, as is visualizability. The strong and routine influence of each of these considerations on theory choice was explained epistemically in the previous section. I am convinced that the other aesthetic properties known to exert similar influence in science—simplicity, beauty, elegance, and symmetry—maintain their position of influence for precisely those epistemic reasons which would lead us to expect familiarity and visualizability to affect the direction of scientific inquiry. Each of these properties plays a comparable role in reducing the amount of cognitive effort required to solve a given problem.

The general conclusion for which I will argue is that scientists appeal to aesthetic criteria in theory choice because those criteria select for frames that demand the lowest amount of cognitive effort required to use them effectively in solving scientific problems. These aesthetic properties need not make a theory more likely to be true to be epistemically important to science. Rather, they make a theory more likely to solve problems, because they make a theory easier to use and thus make the cognitive ends for which the theory is intended easier to achieve.

A theory appeals to scientists' aesthetic sensibilities when it allows them to impose some sort of structure on a set of problems that makes them easier to solve. Part of the cognitive value of structure lies, in Poincaré's words, in how it "enables us to obtain a clear comprehension of the whole," and we can understand the appeal of aesthetic properties generally by appreciating how each one imposes structure on problems in a way that facilitates the synoptic grasp to which Poincaré refers. Familiarity, as we have seen, does this by converting structurally foreign problems into the sorts of problems for which successful approaches are already firmly in hand. Visualizibility does this by allowing scientists to apply geometrical structures and intuitive causal imagery to problems, availing them of powerful nonpropositional forms of reasoning. I now turn to two other prominent aesthetic properties known to be important in scientific practice: simplicity and symmetry.

What distinguishes simplicity from familiarity, and what is epistemically special about simplicity, is that *simplicity achieves in an instant that which familiarity requires past experience to accomplish.* This distinction generalizes to other aesthetic virtues, and it provides a general explanation for the important role of aesthetic virtues in scientific practice. We know from empirical studies that unfamiliar but highly apt metaphors are processed with the same speed as familiar metaphors (Blasko and Connine 1993). This indicates that processing speed is "modulated by aptness" (Bowdle and Gentner 2005, 204). Now, as I argued earlier, aptness judgments in rational inquiry are partly constituted by the judgment that a frame structures research problems into a "convenient and manageable form." As I show later, these aesthetic properties are valued because of their contribution to "convenience and manageability." As such, they contribute to the *aptness* of a research frame.

8.3.1. Simplicity

Simplicity is perhaps the most famous of the aesthetic appealing properties of theories, and philosophers have spared no ink in an effort to explain what it is about simplicity that makes it such an attractive thing for theories to possess. This vast literature orbits around the twin challenges of (1) describing what simplicity is and (2) describing its epistemic significance. Although it is not difficult to find instances of scientists praising theories for their simplicity, as a theoretical virtue simplicity has stubbornly resisted satisfying description. A big part of the problem here has been that we can identify several distinct kinds of theoretical simplicity, each of which has at one time or another been credited as the source of theoretical virtue. To give a symptomatic example, the genus of simplicity criteria that govern scientists' preferences for certain polynomials involves at least three distinct species, since we can describe the simplicity of a polynomial by reference to (1) its degree; (2) whether its terms' exponents are integers; and (3) whether its terms' coefficients are integers (McAllister 1996, chapter 7). An entirely distinct genus of simplicity criteria governs preferences regarding a theory's ontology. We can distinguish distinct species of simplicity within this genus as well, such as (1) the preference that entities not be multiplied beyond necessity (Occam's Razor); (2) the preference for directly observable entities over indirectly observable or unobservable entities; and (3) the preference for physically basic entities over composites. A distinct but related set of factors that poses serious difficulty for attempts to unify the disparate species of simplicity is the fact that simplicity judgments themselves seem to be highly subjective and are, moreover, conditioned by familiarity.[8] In addition to contributing to the proliferation of simplicity types, the element of subjectivity threatens to derail any attempt to unify instances of simplicity under one substantive conceptual umbrella.

Unsurprisingly, simplicity's epistemic significance has been just as contentious as its content. Some of the challenges involved in articulating an epistemic rationale for simplicity's important role in the choice between scientific theories derive directly from the heterogeneity of its content—it is, for example, hard to explain how directly observable entities and integral exponents could satisfy the same epistemic demand. But, as with other aesthetic criteria, the principal challenge has been to describe how simplicity of any sort could contribute to the *well-foundedness* of a scientific inference. The dominant approach to this problem is to argue that a theory's simplicity (however defined) varies directly with probability of its truth. Thus, for truth-seeking reasons, we ought to prefer theories to be as simple as possible.

The many and varied attempts to link simplicity to truth on general principle have not resulted in any perceptible consensus among philosophers. As a rough, unscientific generalization, it is probably safe to say that both scientists and philosophers of science are less convinced of the existence of a simplicity-truth link than any of us were before efforts to uncover such a link began to accelerate during the first half of the twentieth century.[9] And it is precisely

---

[8] McAllister (1996), who also references Harré (1960) and Priest (1976).
[9] For a dissenting opinion, see Schindler (2018).

the proliferation of attempts that has been most compelling in this regard. No account has been able to attract a significant number of adherents. As Kuhn himself pointed out "Postscript," perhaps this is because, as a value judgment, simplicity means different things to different practitioners. Whether we decide to continue with this line of inquiry, the aggressive pursuit of alternatives at this time would seem prudent.

In line with the preceding treatments of aesthetic criteria, I will argue that the pursuit of simplicity in scientific inquiry is motivated by its ability to provide structure to scientific problems in a way that affords an appreciable reduction in cognitive effort. The idea is not entirely new. Quine counts simplicity as one of the two best conditions for "effective thinking"—the other being, as we've seen, familiarity. A similar point was made even earlier by Reichenbach, whose notion of "descriptive simplicity" refers to the kind of simplicity that is "justified by convenience" and contributes to effective thinking or, as he called it, "economy" (Reichenbach 1938, 376). While Reichenbach saw a definite role of descriptive simplicity considerations in theory choice, he made some effort to distinguish its role from that of "inductive simplicity," the latter of which he took to be epistemically significant because of its role in generating predictions. I believe this was a mistake, in the sense that what I take to be epistemically significant about simplicity is precisely its role in generating "convenience" (recall Lorentz's attempt to adjudicate between Schrödinger and Heisenberg in these terms). In addition to fitting elegantly under the same epistemic rubric as the other aesthetic criteria surveyed earlier, adopting this view on how simplicity makes its contribution to well-founded inference allows us to deal with the problems of subjectivity and heterogeneity that have hitherto hampered our efforts to characterize simplicity. Individuals should be expected to differ to some degree with respect to what they find cognitively convenient. If the desideratum of simplicity is justified by considerations of convenience, we should therefore expect individuals to differ with respect to what they regard as constitutive of simplicity. And if individuals differ in this latter respect, we should expect significant diversity in the forms of simplicity that are epistemically influential. Rather than being grounds for thinking that there is no such thing as simplicity that plays an epistemically important role in science, diversity is the expected result given this understanding of simplicity's epistemic function.

Let us begin with the notion of effective use, discussed in conjunction with familiarity earlier in this chapter. There, we saw how familiarization with a particular tool or technique reduces the cognitive effort required to achieve our goals. How was it able to do this? In Kuhn's example of problem sets, the benefits of familiarity came in the form of being able to perceive the way in which each problem can be represented as an instantiation of a general pattern or problem structure; at that point, "only manipulative difficulties remain." The process of familiarization, then, is characterized by a transition in the way we think about these problems. At first, each problem consists of a number of distinct components whose specific relations of relevance to one another must be worked out on a case-by-case basis. But as the familiarization process evolves toward genuine competence, this unordered set of discrete components is converted into a single monadic structure which the proficient deploys all at once, typically without thinking about how it is to be applied. Indeed, the nature of the case is such that the true proficient is often incapable of describing the now-familiar skill in terms of the discrete components which had previously been constitutive of the skill itself; the components have ceased to be cognitively accessible to him (Polanyi 1958).

This same process is well-known to anyone who has formally studied a second language and gone on to achieve fluency. Sentence construction in the beginning stages of language study is a cognitively exhausting process, because, for the beginning student, the new language is a dizzyingly complex system of ordering rules (the grammar) applied to a colossal set of components (the lexicon).

Fast-forwarding to the point of fluency, the system of rules is no longer something the student needs to consult consciously when constructing a sentence. Depending on the duration of fluency, he may no longer even be able to describe the rules, despite observing them with far more fidelity than he did in his early days when he could have recited the rules from memory. Similarly, the lexicon of the new language will now be as basic to him as his native language, and the cognitively demanding task of mentally searching for translations of words will cease to be part of the sentence construction process. At this point, sentences in the new language will express complete thoughts, in the same way they previously had done in his native tongue.

A convenient starting point for illustrating the distinction between familiarity and simplicity comes from Reichenbach's discussion of the way in which our preference for simple equations is "justified by convenience," where he states that "[w]e prefer . . . a simpler analytical expression because we know better how to handle it in a mathematical context" (Reichenbach 1938, 380). In other words, we know better how to *effectively use* simpler analytical expressions in a mathematical context. Our ability to use simpler expressions more effectively does not depend on our experience using those expressions or, indeed, *any* expressions. Nor does it depend on the particulars of the mathematical context in which we are using them. We just find simple analytical expressions easier to cognitively manage and apply, regardless of context or experience. The use of simpler expressions is convenient because (1) like familiar expressions, we already know how to use them effectively, and (2) *unlike* familiar expressions, our ability to use them effectively does not depend on special conditions.

The epistemic justification of scientists' preference for simplicity is closely related to that of familiarity. Because a theory's simplicity makes it relatively easy to cognitively manage and apply, "we know better how to handle it in a mathematical context," whether or not we've encountered that context before. As with familiarity, this ability is epistemically significant because it increases our propensity to generate families of solvable problems. We know that choosing a theory for its simplicity will eventually allow us to adapt it to hitherto undiscovered problems, because our ability to use it effectively is relatively context-independent. This, in particular, is, in my estimation, why simplicity is routinely distinguished from familiarity and singled out as being an especially attractive quality of theories. Simple theories need no track record to recommend them, nor are they opportunistic in the way that familiar theories must be. We can use our possession of a simple theory to project future problem-solving success because a simple theory's success has more to do with what we are good at using than it does with what nature is like. In short, simplicity does not make a theory more likely to be true; it makes us better able to use the theory. And if we are better able to use a theory, there is a greater chance of its contributing to problem solving.

Within this alternative epistemic explanatory framework, we can see how a preference for simple theories can causally promote the success of science without making theories more likely to be true. But that is not all it affords us. We can simultaneously offer an elegant solution to the problems of diversity and subjectivity that have obstructed efforts to characterize simplicity. If the preference for simplicity is propagated by simplicity's role in making theories easier to use regardless of context, we would expect the specific form that the preference for simplicity takes for a given researcher to depend largely on what kinds of features make it relatively easy for her to use a theory no matter what the context. In this sense, what makes a theory "simpler" will be a largely subjective matter. This is not to say that there will be no uniformity across individuals with respect to what form their simplicity preferences take. Shared simplicity preferences can be understood to reflect general features of human cognition. But, of course, we also expect individuals to differ cognitively, including with respect to what sorts of theories they find cognitively demanding.

Moreover, we ought not to have any expectation that what makes things simpler in one discipline, or for one kind of problem, will also make things simpler in other disciplines or for other kinds of problems. The subjectivity of simplicity preferences is not merely apparent; it is real, because individuals can differ in what they find cognitively convenient—that is, in what sorts of features of theories reduce the demand on their cognitive labor. And if the cognitive idiosyncrasies and divergent problem-solving demands of individuals are partly accountable for the specific form that simplicity preferences will take, we can expect there to be lots of variation in the sorts of features that are described as making a theory "simpler," and we can expect consequent variation in which theories are taken to be simpler.

The ability of this alternative approach to simultaneously explain (1) how simplicity preferences causally contribute to the success of science; (2) the apparent subjectivity of the simplicity criterion; and (3) the apparently diverse forms of simplicity preferences is not something that can be claimed by the "simplicity-truth" view. Indeed, both the diversity and subjectivity of simplicity criteria seem bizarre from that view's perspective. With respect to diversity: if simplicity is a guide to truth, then either (a) by some astonishing coincidence, the many ways in which a theory can legitimately be described as "simple" all bring us closer to the truth, or (b) the apparent diversity of simplicity criteria is illusory; there is something substantive which all the diverse manifestations of the simplicity criterion share and which is connected to truth. With respect to subjectivity: if simplicity is a guide to truth, then (a) by an astonishing coincidence, our (diverse) subjective preferences are somehow tuned into the truth, or (b) the apparent subjectivity of simplicity criteria is illusory.

By contrast, the alternative, cognitively oriented view I've argued for is able to validate the apparent subjectivity and the apparent diversity of simplicity preferences as real *and* as epistemically significant. This is an important advantage over the "simplicity-truth" view, in that (1) it is able to explain these appearances on their own terms, rather than having to explain them away as illusory; and (2) it is able to explain both appearances via the same mechanism it uses to explain simplicity's epistemic significance. As to (1), it is customary to prefer explanations that treat appearances as real over those that need to explain away appearances as illusory, other things being equal. But the premium for explaining away the appearance of subjectivity would seem even higher in the case of something like a preference for simplicity, because of that sort of preference's long and direct association with aesthetic judgment, which is paradigmatically subjective. If simplicity judgments are a form of aesthetic judgment, we would *hope* that our explanation of their presence in scientific inquiry to have a lot to do with the subjective preferences of individual scientists.

8.3.2. Symmetry

Other aesthetic properties contribute to problem structuring in their own way. The property of symmetry, which is routinely held up as a highly desirable element of theories in modern physics, is intimately associated with the notion of structure. At its core, the modern understanding of symmetry is in terms of interchangeability of parts. For example, to say that your face is perfectly (vertically) symmetrical means that we could take a piece of the left side of your face and interchange it with a piece of the right side of your face without changing the way your face looked in any way. This symmetry can be thought of as a mathematical rule that describes a certain relationship between the points that make up your face— namely, that for every pair of coordinates (x,y) that make up your face, there is an equivalent pair $(-x,y)$, such that we can transform the x-values for your face-points from positive to negative (and vice versa) without your face undergoing any change in and of itself. In mathematical terms, we say that your face is *invariant across such transformations*.

For anything that possesses a symmetry, its symmetry can be described in terms of the set of transformations across which that thing remains unchanged. The symmetry of your perfect face is described by one such transformation. Other objects, like some ideal geometrical figures, have more symmetries, and so the set of transformations across which they remain unchanged is larger than the one that describes your face's vertical symmetry. Squares, for example, are invariant across eight different sorts of transformations: all rotations of 90°, 180°, 270°, and 360°; vertical reflection (like your face); horizontal reflection; and reflection across each diagonal. Any such set of transformations that leaves an object unchanged is called a "symmetry group."

There are various kinds of symmetry groups, each representing a distinctive mathematical structure. Squares belong to a symmetry group named D_4 ("D" for "dihedral" [i.e., regular polygon], "4" for the number of axes of reflection). This group also contains the regular star polygons {12/3} and {8/2}, also known as the Jerusalem Star. In fact, each regular polygon belongs to one of the D_n symmetry groups. We can use our knowledge of the mathematical properties of a given symmetry group to study its known members in much greater detail and to identify new group members. In general, we can use the mathematics of groups to study anything with symmetry, because "[w]henever symmetry occurs, groups describe it" (Fletcher 1966, 399).

Of course, symmetry has been a highly valued property of scientific theories long before we came into possession of a sophisticated mathematics for describing it. But that does not mean that the appeal of symmetry must therefore be independent of its relation to these well-defined group structures. At a basic, nonmathematical level, a symmetry gives us direct insight into the range of changes or counterfactual conditions that can be ignored without having any effect on our "clear comprehension of the whole." For example, since we know your perfect face is vertically symmetrical, we know that it makes no difference whether we're looking at you face to perfect face, or a reflection of you; it is exactly the same image in both cases. For squares, the range of completely irrelevant changes consists of all the transformations in the D_4 symmetry group.

The cognitive freedom that symmetry affords us is a plausible explanation for why it is a highly desirable quality of scientific theories. Evolutionary biology is not normally known for its theoretical symmetries, but an episode in the career of Stephen Jay Gould provide a particularly instructive illustration of how symmetries structure scientific inquiry to increase the solvability of problems. As described in Chapter 5, Gould spent much of his scientific career trying to lift paleontology out of what he regarded as a scientifically primitive state of merely describing and cataloguing fossils to a more modern phase of inquiry which employed the use of models to test theories about the history of life. His strategy for achieving this was designed around a commitment to searching for evolutionary processes that were symmetric with respect to taxa.[10] As suggested, this example highlights the direct connection between scientists' preference for symmetry, on the one hand, and symmetry's ability to specify what sorts of changes leave our "clear comprehension of the whole" intact, on the other.

His first scientific papers examined the use of a simple mathematical model for the study of fossil trends involving coordinated changes in size and shape within a lineage: a power function $y = bx^k$, known to biologists as the "equation of simple allometry." In plain terms, the equation of simple allometry states that body size and body shape change together in simple geometric proportions. This is because, as a matter of simple geometry, "certain shape alterations are mechanically required by size increases" (Gould 1966, 588). A number of biological functions require specific area-to-volume

---

[10] See Gould (1970) for the first general statement of this commitment. See Haufe (2015) for other examples from Gould's oeuvre.

ratios. When there is selection of increased body size, there will be an associated increase in surface area. But size increases as the cube of linear dimensions, whereas surface area increases as the square of linear dimensions. Organisms must therefore alter their shapes to preserve their adaptive area-to-volume ratios.

In 1966, Gould penned a lengthy review essay in which he praised the equation of simple allometry for its "long history of fruitful use" in biology, which at the time went back at least seventy years to its role in the quantitative description of brain-weight to body-weight relationships. One of his goals in that essay was to explain why biologists were "almost exclusive" in their fidelity to the power function as the preferred means with which to study biological correlations. Here Gould observed that one virtue the power function did *not* have over other methods was superior accuracy of fit with empirical measurements, and he noted a number of alternative techniques what had "been applied to obtain closer fits to point scatters" (Gould 1966, 596). While most of these techniques were more complex than the power function, it was also not the case that its advantage lay in its greater simplicity. No less a personage than D'Arcy Thompson "was unimpressed" (Gould 1966, 595) with the power function "and showed that data which had been assumed to show the allometric relationship were equally well described by simple linearity" (Sholl 1954, 229). In addition, it had been known for at least a few generations of biologists that the correct mathematical description of organismal growth was not going to be something which we could even achieve, much less use:

We are now more fully aware of our inability to specify the many factors that may be responsible for growth in terms of a few parameters, let alone finding a mathematical statement about their relationships; in any case, such a relationship would be of such a complexity that it would not be expressible in terms of simple mathematical functions. (Sholl 1954, 225; cited in Gould 1966)

Since *a fortiori* the power function was known to not be the correct mathematical description of growth, why then was it able to command the allegiance of almost everyone in the discipline?

The principal virtue of the equation was that it was, in Sholl's words, "highly adaptable." This virtue rested on the symmetry assumption at the heart of equation—viz., that the morphological constraint of geometric proportionality was invariant across taxa: no matter what species we study, and no matter what period of time we're looking at, morphological changes can be expected to obey these geometric constraints. In this way, the equation of simple allometry had allowed biologists to step back from the complex, heterogenous, and generally unavailable biological details concerning developmental processes—details that are expected to be highly variable both within and across taxa—and structure their inquiries in such a way as to make the whole mess effectively irrelevant. Had that *T. rex* been a butterfly, its body size still would have changed as the cube of linear dimensions, while its surface area changed as the square of linear dimensions. And thus, its body shape would have had to change in predictable ways to preserve its particular area-to-volume ratio. By using a model that was symmetric with respect to taxa, biologists were able to maintain a "clear comprehension of the whole"—that is, an insight into fundamental aspects of an important general process—and avoid the blurring of their cognitive vision with an overabundance of detail.

Besides illustrating the epistemic contribution of aesthetic criteria, both the quantum mechanics and Gould episodes shed light on how the acceptance of a particular approach to understanding nature can be explained as epistemically rational even if the motivation for that acceptance is largely aesthetic. For the scientists involved in these episodes, the satisfaction of aesthetic criteria was reliable evidence of an approach's potential fruitfulness, *not* because those aesthetic criteria are correlated with truth in some way, but because they knew that approaches which satisfy those criteria are easier to use. They were able to project the future success of the approach at these early stages largely because they were able to project their own cognitive proclivities. This, in general, is how new approaches can be rationally embraced by the scientific community—or sizable portions thereof—in advance of sustained empirical inquiry and in advance of an accumulated record of solved problems. In certain instances, the basic *structure* of the theory satisfies certain aesthetic criteria. The mere satisfaction of those criteria tells us something epistemically important about the theory's "future promise," because satisfying those criteria reduces the amount of cognitive effort required to solve problems.

Of course, this is not to deny the epistemic significance of empirical criteria like adequacy and novel predictive success. Rather, our aim is to expand the store of genuinely epistemic motivations for adopting a theory, by showing the variety of factors involved in producing solved problems. Empirical data and experimental testing are unquestionably important factors. But I take the evidence presented herein to indicate that a theory's ability to structure problems in a way that makes it easier to see what sort of data are relevant (as well as irrelevant) to their solution, or that makes it easier to treat them mathematically, is every bit as critical for generating scientific achievement; perhaps it is of far greater importance. Structuring problems in convenient and manageable ways leads eventually to their being solved. Its more immediate—and, perhaps, more valuable—effect on scientific inquiry, though, is to generate classes of problems that *can* eventually be solved. This effect is what I have called *fruitfulness.*

9

Close Shave?

Since the second half of the twentieth century, historians of science have moved away from historical explanations of problem solving[1] and toward depictions of scientists' views as highly contingent products of time and place, paying increasing attention to the possible role that incidental features of context such as social norms, political expediency, and even diet might play in substantively shaping the nature and direction of scientific inquiry at a given time. The effect—sometimes intended, sometimes perhaps not—has been to "lower the tone in the history of science," adopting the position that science is "not cognitively or methodologically or socially unique" and showing how it too is "heterogeneous, historically situated, embodied, and thoroughly human." Indeed, Steven Shapin has declared this approach to be "what now counts as the history of science" (Shapin 2010, 14).

The threat to scientific knowledge presented by highly contingent events in the history of science can be modeled as a species of violation of a necessary condition on knowledge itself, what Collins (2006) has dubbed the Close Shave principle. This principle identifies as epistemically problematic instances in which we very easily might not have held the beliefs we actually do have and which we take to constitute our current knowledge. The "tone-lowering" emphasis on Close Shaves standardly appears in three varieties: (1) accidental discovery; (2) accidental theory; and (3) accidental consensus—each of which, by violating the Close Shave

---

[1] See Laudan (1993).

Fruitfulness. Chris Haufe, Oxford University Press. © Oxford University Press 2024.
DOI: 10.1093/oso/9780197666395.003.0009

principle, poses a special challenge to a defense of science as cognitively authoritative.

Consider: there is a possibility very close to actuality in which Roentgen did not discover the x-ray; by a close shave, he did. This very familiar species of Close Shave in science is the allegedly serendipitous nature of scientific discovery, a legend which makes for good popular narratives but bad arguments in favor of basic research funding from Congress,[2] as well as from science funding agencies. What might have appeared to the scientific community to be a good strategy for getting the public excited about science[3] may have had the unintended consequence of undermining part of the rational basis for trusting science. It is not difficult to understand why these institutions would be reluctant to invest in a vision of future science whose success hangs on the appearance of the Unforeseen. Nor are the epistemic consequences of an inherently accidental discovery process difficult to apprehend: what is so impressive about scientific knowledge if most of the important scientific discoveries might just as easily have failed to take place? Shouldn't we expect something more robust from our best knowledge-generating practices?

Another sort of Close Shave focuses on how "external" influences might have affected the development, content, and/or interpretation of a particular scientific theory or method. For example, it has often been claimed that the causal significance that Darwin attributes to intraspecific competition is on account of his having come out of the highly competitive trappings of Victorian England (e.g., Todes 1989). These analyses aim to represent the appearance of a particular scientific theory as a historical accident that depended on the highly fortuitous confluence of factors not directly related to problem solving: there is a possibility very close to

---

[2] See the many examples documented in Greenberg (2000).
[3] See again Greenberg (2000) for many examples.

actuality in which Darwin did not formulate a theory centered on intraspecific competition; by a close shave, he did.

Yet another sort of Close Shave attack attempts to portray certain historical episodes of scientific consensus as similarly incidental to any concerns related to the well-foundedness of knowledge, conditioned instead on mere accident, on personality, or on certain prevailing sociocultural, political, or economic norms. Perhaps there is a possibility very close to actuality in which Bohr's indeterministic quantum mechanics was not the favorite at the 1927 Solvay conference, losing out to de Broglie's deterministic version; by a close shave, Bohr carried the day, and quantum mechanics consequently came to be a symbol of the fundamentally incomprehensible predicament that is Nature, instead of a well-behaved classical system. Shapin and Schaffer (1985) famously argued that the experimental philosophy championed by Robert Boyle and other members of the early Royal Society was ultimately taken up by the burgeoning community of professional natural philosophers because it accorded well with certain political ideals that held sway in England at the time. According to this view, even if the development of the experimental philosophy is explicable purely in terms of effective problem solving, its subsequent widespread popularity can most plausibly be accounted for by noting political expediency.

Had any of these possibilities been actualized, the history of science would have developed along an entirely different trajectory, and (the argument goes) we would not have adopted the beliefs about nature that we currently hold. These ideas seem to lack an important kind of stability. Therefore, none of them qualifies as scientific knowledge in the epistemically privileged sense. Now, because notions like *progress*, *understanding*, and *discovery* are so intimately bound up with the concept of knowledge, a track record of violating necessary conditions on knowledge destabilizes the track record's ability to serve as a signal of science's epistemic power. For, if the history of science is a history of Close Shaves, then either (1) science does not actually produce epistemically special phenomena like progress, understanding, and discovery, or (2) science *does* produce those phenomena—they're just not epistemically special. In the latter case we will need to revise our understanding of them in ways that involve divorcing them conceptually from improvements in the well-foundedness of our knowledge, as social constructivists have attempted to do.

It is not obvious from the perspective of epistemology how we should approach this now-dominant strain of scholarship in history of science. One thing that *is* obvious is that the threat of Close Shaves to the cognitive authority of science must be diffused: if the historical development of science is a history of Close Shaves, then there is no discernible reason to invest science with the cognitive authority it routinely demands. For, if the history of science might easily have been otherwise, then scientists' views might have easily been otherwise. If that's true, then it is difficult to see why we should take science's *actual* claims about nature as authoritative.

The epistemically toxic effect of Close Shaves suggests that whatever is causally responsible for epistemically special phenomena like scientific progress should be an intended consequence of widespread principles of rational inquiry. This is not to say that scientists need to be able to foresee every event that might arise in the development of some line of inquiry for scientific progress to be preserved as the product of a rational process. But progress itself cannot hang delicately upon a series of Close Shaves and still appear to be a predictable consequence of the systematic attempt engage with nature that we take science to be. Rather, progress must be understood to flow naturally from the way in which scientists characteristically make choices in the context of inquiry and the ways in which scientific inquiry is organized. If discoveries are largely happy accidents and if individual scientists are subliminally lured into sympathy for certain theories, the cognitive authority of science is in jeopardy.

A Close Shave in the history of science is a case where a particular historical instance of scientific discovery, theory formulation, or consensus seems as if it could easily have been otherwise, owing to the possible causal influence of historically contingent factors of time and place. Had these historically contingent factors been otherwise, a certain scientific discovery might never have been made, a certain scientific theory might have been vastly different, or the scientific community might never have converged on the particular theory or methodology it ended up adopting.

The fact that there is a close possible world in which established scientific knowledge is very different is epistemically significant because it suggests that there is a substantial element of luck involved in the historical development of science. The more luck there is associated with a belief, the more fragile my possession of that belief appears to be. The more fragile my possession of a belief, the less it appears to qualify as genuine knowledge. My correct belief about next week's lottery numbers is not knowledge; it is a lucky guess. Since my belief about what the lottery numbers will be is not affected in any way by what they will be, the coincidental connection between my belief and the facts is extremely fragile; there is no reason to suppose the connection would have held had the lottery numbers been different or had my belief been different.

If the historical development of science involves a substantial amount of luck, then scientific knowledge possesses this same fragility. The constituents of what we call "scientific knowledge" could, owing to historical contingency, easily have been otherwise. If contingency begets fragility, then there is a close possible world, nearly identical to ours, except that scientific knowledge is composed of a vastly different set of commitments—different discoveries, different theories, different methodologies. And if our commitments might easily have been otherwise, owing perhaps to differences in cultural circumstances, then there's no reason to invest our current commitments with any epistemic privilege. Had we possessed different commitments, we would have been as confident in those as we are in our actual commitments. But if we regard the commitments we might have possessed as epistemically deficient in some way—being as they are at odds with our current commitments—then we ought to regard our current commitments as equally deficient. After all, the only thing separating them is a slight difference in cultural circumstances.

I want to examine a number of strategies for reducing the threat level that Close Shaves present for a defense of the cognitive authority of science. It will turn out that our ability to successfully implement the most promising of these strategies is affected by our conception of which paradigmatically epistemic goal scientific inquiry is oriented toward. Most importantly, our success in this regard will depend largely on whether we can articulate a goal-oriented process which buffers scientific knowledge against the threat of Close Shaves. Some paradigmatically epistemic goals are better suited than others for this task.

9.1. How to Avoid a Close Shave

There are two very general strategies for inoculating a particular bit of scientific knowledge against demotion to "Close Shave" status, both of which can be applied with equal effectiveness in each of the domains where Close Shave claims are traditionally made—(a) discovery, (b) formulation or interpretation of theory, and (c) consensus. The first general strategy involves building a case for *plausible deniability of causal influence* of the kind that is frequently asserted—namely, the influence that factors having nothing to do with improving the well-foundedness of inference might have played in shaping scientists' decisions in the context of inquiry. This approach takes a cautious, measured skeptical perspective on such claims. The second general strategy accepts that these sorts of factors have had a major impact at given level of inquiry, but *posits the existence of higher-level features of rational inquiry* which prevent that lower-level feature from significantly affecting the overall trajectory of inquiry—in much the same way that a single, mostly impotent genetic mutation makes no effective difference to a large population's evolutionary trajectory. It may make some difference to the life of an individual, but not to the larger evolutionary process of which he is part. So it is, I argue, in science: one level's Close Shave may just be another level's epistemically insignificant variation.

I discuss both of these strategies, weighing their potential effectiveness and, ultimately, attempting to outline in more detail the demands on an account of rational inquiry that is capable of taking seriously the historical contingency of scientific inquiry while simultaneously underwriting the cognitive authority of science.

9.1.1. Plausibly Denying Causal Influence

There are a number of ways we might endeavor to make sense of claims concerning the influence of details of historical context on the direction of scientists' views or on the direction of scientific inquiry more generally. One broad set of strategies would involve stronger and weaker versions of denying what has been asserted. A strong version would deny the accuracy of historians' descriptive claims about the composition of a given historical context—for example, that Darwin's Victorian society was really all that competitive. Unless one is willing and able to get down in the historical muck to refute these descriptions, this approach to defusing the epistemic threat posed by historical contingency looks like a nonstarter. A good rule of thumb in scientific inference is: *if data conflicts with well-confirmed theory, trust well-confirmed theory.* Since there are no well-confirmed philosophical theories of science, we can't justifiably trust some well-confirmed theory against the data that historians have amassed.

A weaker version of denial admits that "external" factors have had a causal influence on particular theories, but not on the aspects of those theories that are responsible for the theories' success.

Again, using Darwin as our example, we might admit that the reason Darwin's particular version of the theory of natural selection places a lot of emphasis on competition is because Darwin's society was itself very competitive, but aver nevertheless that the concept of competition does not bear an important portion of the explanatory load in account for the theory's success. This weaker version of denying causal influence really depends on the historical facts concerning (a) which bits of a theory are responsible for its success, and (b) how those bits came to be incorporated into the theory in the first place. Both species of fact are, of course, difficult to come by. In the case of Darwin's theory of natural selection, debates over which aspects of the theory are responsible for its success are ongoing and the views on this matter are diverse. Competition, for example, is now considered to be a special case of a more general class of means by which reproductive success differentials can be achieved. And reproductive success differentials are now thought by many to be just a special case of a more general class of means by which probability distributions shift over time (Shimony 1989; McShea and Brandon 2010).

But even supposing that we achieved consensus on the causal importance of the concept of competition to the theory's success, that would still leave the problem of discerning how the concept found its way into Darwin's theory. This too is a delicate issue. We know that Darwin's reading of Malthus in September 1838 was the critical point in this regard (Ospovat 1981). But that does not resolve the matter, because to motivate the claim that Darwin's competitive context actually made the concept of competition appear to him most suitable for explaining biological adaptation, we would need evidence attesting to the idea that *Malthus's work would not have resonated with Darwin, had Darwin's society not been so competitive*. Ideally, we should also have evidence that Malthus's work similarly influenced philosophical naturalists in similarly competitive societies. Now, we do have evidence that Darwin's Malthusian framing of the theory fell on deaf ears in less socially competitive Russia, even among those who otherwise accepted natural selection as the principal mechanism for evolutionary change (Todes 1989). In addition, Alfred Russel Wallace grew up in the same place as Darwin, and he also claimed to have been influenced by Malthus in his theorizing (Slotten 2004, 144).

This is all very suggestive. But it's also worth recalling that Darwin's and Wallace's shared background included more than membership in Victorian society. In particular, they were both eminent field naturalists, with a special knowledge of tropical ecosystems—Darwin's in the West, Wallace's in the East—which are characteristically far more densely populated than those of other latitudes. Out of all the Victorian Englishman who read Malthus's *Essay on the Principle of Population* in the sixty years between its publication and the discovery of the principle of natural selection, only one person besides Darwin was similarly moved by it, and he happened also to have had the same "seminal field experiences in densely populated tropical environments" that formed the empirical background of Wallace's theorizing (Todes 2009, 36). In addition, Todes (2009) notes that Russian naturalists' *field* experience—not their social ideals—told against a Malthusian framing of natural selection: out on the Siberian steppe, the struggle for existence was not intraspecific, but rather against the extreme abiotic conditions for which that region is well known. Considered alongside this alternative explanation, the case for Victorian influence seems somewhat less compelling.

The influence of Malthus on Darwin is one of the most well-studied instances of theory construction in the historical literature on science. And yet, even here, we still cannot answer as simple a question as whether Malthus influenced Darwin because of his social background or because of his scientific background, or what the relative significance of either was—that is, not without a lot of additional information of the sort I've alluded to earlier. All this is to say that knowing whether the "external" factors have had the significant influence they are often claimed to have had in the formation of scientific theories requires us to be able to disambiguate candidate influences with confidence on the basis of the historical record. There's just no other way to say it: this is very, very hard to do. Because of the significant challenges involved in building a convincing case, we are right to be cautious when confronted with historical claims about causal influence. Motivating a case for causal influence, after all, must involve more than vague coincidence and innuendo. And the evidentiary burden is particularly high when something as important as the cognitive authority of science is on the line.

Although there is a strong case for insisting on a higher standard of evidence when it comes to claims about the causal influence of "external" factors on the direction of scientific inquiry, it is not hard to imagine that standard being routinely achieved by historians once it begins to be widely observed. Were this higher standard to be routinely achieved in historiographical practice, there would cease to be much basis for denying the causal relevance of social, cultural, historical or otherwise nonepistemic factors for shaping the views and strategies of scientists at a given time. Presumably many members of the Science Studies community believe we've already reached that point. Whether we have or not, we need a strategy for how to preserve the understanding of science as epistemically special once the case for significant causal influence of contextual factors reaches the level of implausible deniability.

In what follows I describe a few versions of a general recipe for how to reconcile Close Shaves with the epistemic integrity of science. The common ingredient to each of these is the existence of higher-level processes of inquiry that are ultimately insensitive to variation at lower levels. I illustrate this recipe in the contexts of discovery, theory formulation, and consensus.

9.1.2. Resisting Close Shaves in the History of Scientific Discovery

The historian of science Frederic Holmes spent his distinguished career mapping out the "investigative pathways" along which a number of notable scientists traveled throughout the course of their research and which ultimately resulted in historic scientific discoveries. At the unfortunately early end of a remarkable effort, Holmes (2004) looked back on the careers of the scientists he and others had studied in search of any patterns that could be discerned through a side-by-side comparison of their individual investigative pathways. He found several.

The existence of patterns that are common across discovery-prone investigative pathways is significant because it hints at the possible existence of higher-level processes that causally explain those patterns. In this way, such patterns allow us to see a particular individual's significant discovery as more than a serendipitous event in what would otherwise have been a lackluster scientific career. Rather, they suggest that this individual is following a well-trodden causal trajectory, one which reliably generates significant discoveries and which thus can potentially feature in an explanation of why science is epistemically special. Thus, it might be true that, but for a number of fortuitous events, Heinrich Hertz would not have discovered how to produce electromagnetic waves. Nevertheless, on Holmes's analysis, Hertz's research career bears the telltale signs of other research careers that resulted in significant discoveries. So, while his historically actualized discovery of electromagnetic waves may have been a Close Shave in the epistemically problematic sense, his having discovered *something significant* was not a Close Shave: the causal pattern manifested by his research career suggests that a significant discovery of some sort was a predictable outcome of Hertz's research. Had he not discovered electromagnetic wave propagation, he probably would have discovered something else.

If science is genuinely epistemically superior to other forms of inquiry for studying nature, we should be able to explain why it is so much better at discovering stable and significant features of nature than other forms of inquiry. Careful analysis of the investigative pathways that have led to notable discoveries reveals the process of discovery to be a far less mystical and chancy one than it is frequently reported to be. These narratives may be less exciting than the accident-in-the-lab adventure stories to which we're often treated, but they are more easily and more convincingly incorporated into (1) a description of an overarching causal process that predictably terminates in scientific discovery and, ultimately, (2) an argument for the special epistemic status of scientific inquiry. The essence of this process is the presence of a particular individual on an investigative pathway that manifests the symptoms of a discovery-prone research trajectory. The tendency of this process to terminate in significant discovery per se is robust across a range of external influences that may affect which specific discovery ends up being actualized. This approach highlights in a simple and direct way the extent to which scientific progress does not depend on the specific commitments that the scientific community ends up adopting. The carefully reasoned pursuit of fruitfulness per se can promote progress independently of what variety of fruit pops up.

9.1.3. Resisting Close Shaves in the History of Scientific Theorizing

The challenge that external influences present to science's cognitive authority is packaged in a set of assumptions about what would have been otherwise. We have just seen one approach which can, on the one hand, acknowledge the Close Shave status of the historical particulars concerning which significant discovery an individual actually made and, on the other hand, still allow us to preserve the epistemic significance of that discovery by portraying *discovery itself* as the predictable outcome of a certain type of research trajectory. According to this view, an individual's specific discovery might easily have been otherwise, but her having discovered something or other of significance probably would not have been otherwise.

Just as we can attempt to restore the epistemic significance of historically contingent discoveries by appealing to a higher causal power, so too can we use the notion of higher-order processes to protect the epistemic merit of scientific *theorizing* against the vagaries of history. Externalists move from the premise that the content of theory *T* has been influenced by certain social conditions to the conclusion that *T* would not have arisen had those conditions not obtained. The damage to science's authority is inflicted by the implicit assertion that *T*'s place in the pantheon of science is not the result of a reliable epistemic process, but rather of a certain set of historical accidents. Since *T* was a Close Shave, either (1) *T* cannot qualify as scientific knowledge; or (2) *scientific* knowledge is not epistemically special. Any account that aims to preserve the cognitive authority of science while taking the history of science seriously must—as with discovery—endeavor to distinguish between aspects of the historical development of scientific *theory* that might easily have been otherwise and aspects that probably would not have.

One way of responding to this version of the epistemic threat is to insist that the move from "socially influenced" to "would have been otherwise" is illicit because it assumes that there is a 1:1 correspondence between sets of social conditions, on the one hand, and scientific theories, on the other, such that every set of social conditions comes with a set of possible scientific theories that are unique to it. This is not a well-motivated assumption.

First, the many instances of multiple discovery *prima facie* undermine the notion that a given discovery would not have happened in the absence of certain social conditions. If these instances really are explicable by reference to just the right overlap in social conditions between the various contexts of discovery, then that is something that needs to be demonstrated directly, not inferred on the basis of multiple discovery itself. In many cases, it simply strains credulity to imagine that the right sort of overlap in the necessary social conditions exists. Although one cannot prejudge the matter, it would certainly be surprising if it turned out that the same social conditions that led US mathematicians John Tukey and James Cooley to develop an algorithm for discrete Fourier transformations in 1965 were also responsible for its independent formulation 160 years earlier in Germany by Carl Friedrich Gauss (Heidman et al. 1985). Even more surprising would be if we were to find that the thirteenth-century Arab Muslim jurist Ibn al-Nafis and the sixteenth-century English physician William Harvey shared so much of their social context that they were both induced to formulate the same view regarding the function of pulmonary circulation.[4] What is perhaps the most confusing thing of all is why—if the two social contexts were so similar—of all the citizens of thirteenth-century Mamluk Egypt, and of all the people living in sixteenth-century Oxfordshire, the only two people in either of these societies to formulate this particular view about pulmonary circulation were the most accomplished physiologists of their respective generations.

Multiple discovery is not the only difficulty that we can raise for the notion that particular theories and methodologies require very specific antecedent social conditions. Apart from the fact that this just does not seem to be true based on an assessment of the apparently highly disparate social conditions in which theories have repeatedly appeared, there is also the more obvious point that there is simply no compelling reason to think that theories *would* be constrained in the way suggested by the insinuation that a particular theory would not have arisen had certain social conditions not obtained. When we accept the claim that, say, Darwin's theory of natural selection was informed by his Victorian social context,

---

[4] See the interesting book by Fancy (2013) for a discussion of the parallels.

we do not thereby commit ourselves to the view that it would have been impossible for the theory to have been formulated outside that context, or even that it would have been more difficult. While a particular scientist's formulation of a particular theory does in fact follow a specific historical path, the *theory's* formulation needn't do so *necessarily*. The formulation of that theory per se may be a much more robust phenomenon, compatible with a variety of different causal histories. In other words, the formulation of the theory itself need not have been a Close Shave just because a particular historical individual's responsibility for that formulation *was* a Close Shave.

If this is correct, then science's epistemic authority has been immunized against a particular strain of skepticism which focuses on the sensitivity of individual theorists to nonscientific aspects of their time and place, and the role that sensitivity can play in theory construction. As real human beings, theorists *are* sensitive to their time and place in ways that may affect the formulation of theories, methods, and every other element of scientific practice (duly acknowledging the fact that some kinds of scientific disciplines will be more susceptible to this than will others). But the theories themselves cannot thereby be indexed to particular social contexts for their formulation and interpretation. If the content of theories was so intimately connected with the particular social conditions out of which they arise, we should expect the theories to be far less generally intelligible than they typically are. Unsurprisingly, scientific training is the common thread uniting the small group of individuals who are able to formulate, interpret, and apply a particular theory.

This common thread—the same one lurking behind the instances of multiple discovery mentioned earlier—forms part of the basis for explaining how scientific progress can accommodate the causal influence of "external" factors on theorists. While the logical point about the illicit move from *theorists' sensitivity* to *theories' sensitivity* is sufficient to block one sort of attack on the epistemically special status of science, positively motivating the case for epistemic privilege requires that we provide some positive account of the process by which the appearance of certain theories in science or in mathematics *can* be robust across various sets of social, cultural, and historical conditions. In other words, there must be some description of the higher-order mechanism that could plausibly underwrite assertions such as: had Newton not formulated an inverse-square law to account for the elliptical path traced by the planets around the sun, someone else would have.

9.1.4. Resisting Close Shaves in the History of Scientific Consensus

We've just seen a strategy for how to reconcile historical contingency with scientific progress that attempts to block the assertion that particular theories are Close Shaves, in the sense that their appearance would have been otherwise but for the appearance of certain background social conditions. This is not the only option available to us. Another version of this strategy seeks to preserve the epistemic significance of scientific consensus by, on the one hand, accepting the historically contingent fact that consensus happened to form around a particular theory or framework, while, on the other hand, denying that the epistemic significance of scientific consensus is rooted in the particular theory around which consensus forms. In the same way we suggested earlier that the appearance of particular theories might be robust across a range of different social conditions, so too might it be the case that the degree of scientific progress made possible by scientific consensus is robust across a range of radically different theories, such that we would expect that same degree of progress if consensus had formed around an entirely different theory than the one it actually did.

What grounds do we have for suspecting that the degree of scientific progress made possible by consensus may be insensitive to certain kinds of variation in the specific theories around which consensus forms? Once again, the case comes down to the existence of symptoms of a higher-order causal process. The fact is that subcommunities of scientists regularly agree on a narrow set of problems that are of special importance, set out to solve them, and eventually succeed in doing so. We can plausibly assert that scientific progress need not have been a Close Shave merely because the specific theories that happened to host that progress *were* Close Shaves. The fact that they might easily not have appeared or might easily not have been the subject of scientific consensus need not imperil the general prospects for scientific progress. Drawing on our earlier example, it might be that, had Bohr missed the train to Solvay in 1927, de Broglie's deterministic pilot-wave mechanics would now command the physics community's assent. But it also might not matter to the progress of science. The rate of progress might simply be indifferent to which of these interpretations we pursue.

If this is correct, then we have another means by which the notion of scientific progress can coexist comfortably with a radical contingency about the order and presence of events in the history of science. Unlike the previous approach, which treats *theories themselves* as exhibiting a certain range of invariance with respect to social conditions, this approach depicts progress as the invariant phenomenon, buffered against the peculiarities of theory that blinker in and out as history unfolds. In this way, we can also address the scientific community's *prima facie* epistemically problematic tendency to undergo radical changes in research frameworks. If what matters for progress is simply whether problems are being solved, and not what we're using to solve them, then radical changes in research frameworks are in principle incidental to the overall progressive tendency of organized inquiry. But again, some positive account is in order regarding the higher-level mechanism that can produce progress no matter what theories happen to arise.

9.2.   Apt Frames and the Inevitability of Progress

I have argued that theory choice is fundamentally about judging the aptness of a frame for scientific inquiry, in much the same way that we judge the aptness of a metaphor (albeit subject to different desiderata). But many a metaphor can be used to aptly characterize the same subject. Could not the same be said for science? That is, if all practitioners care about is aptness, why exhibit a preference for one apt frame over another?

One plausible answer is that different framing devices often facilitate inquiry in different ways. Philosophers have long since progressed beyond the notion that the choice between empirically equivalent theories cannot be based in reason. If practitioners find that one frame makes things "convenient and manageable" while another does not (e.g., a linear approximation versus a nonlinear one), the choice in favor of the convenient and manageable frame is in no way arbitrary. For that matter, neither is the choice between apt literary metaphors. No one thinks that a poet's choice of metaphor is uniquely determined. A poem is constructed to, say, elicit a specific emotional response. A poet will select from among a range of apt metaphors whichever one he finds to be most effective at achieving that response.

This brings us to the second reply: all metaphors are not created equal. While there may be a range of apt metaphors, some of them may be *particularly* apt. No one will ever top J. D. Salinger's description of Ethel Merman. Conceiving of electricity as a fluid seems to have been a particularly apt way of characterizing it in the mid-nineteenth century—less so now, although one could certainly do worse than Maxwell's little tubes channeling electrical fluid. It may happen that at a given moment in the historical development of inquiry, some framing device stands out as particularly apt among the set of apt frames. This amounts to nothing more (and nothing less) than saying that one particular frame is recognized as facilitating further inquiry more effectively than any other frame available. Conceived of in this way, it would make no more sense to call this the "correct" frame for inquiry than it would to say that Salinger's description of Ethel Merman was correct. Rather, the mathematicians' word, "natural," seems to be a close analogue (Tappenden 2008).

Although I do not think we can be confident that science will inevitably discover the True Theory, or that it would have formulated (approximately) the same understanding of nature, had historical events been otherwise, I do think we can be confident that communities of rational inquiry of the sort that characterize science and mathematics will consistently develop increasingly effective ways of generating apt characterizations, which then tend to become increasingly widespread. As I have argued elsewhere (Haufe 2022), these communities should be understood as highly effective systems for optimizing and promulgating problem-solving practices. But it is in the nature of such communities that, in a vast range of cases, things might have easily been otherwise. Had Maxwell not proposed his version of a fluid frame for electricity in the mid-nineteenth century, had Raup et al. not proposed "particle paleontology," would someone else have? I can find no reason to think so. However, would their respective communities eventually have adopted a framing device which its members judged to be particularly apt? That, I think, was inevitable. At the very least, the historical trajectory of every branch of modern science attests to its inevitability.

References

Andersen, Kirsti. 1985. "Cavalieri's Method of Indivisibles." *Archive for History of Exact Sciences* 31 (4): 291–367.

Bacciagaluppi, Guido, and Antony Valentini. 2009. *Quantum Theory at the Crossroads*. New York: Cambridge University Press.

Barnosky, Anthony D., Nicholas Matzke, Susumu Tomiya, Guinevere O. U. Wogan, Brian Swartz, Tiago B. Quental, Charles Marshall, Jenny L. McGuire, Emily L. Lindsey, Kaitlin C. Maguire, Ben Mersey, and Elizabeth A. Ferrer. 2011. "Has the Earth's Sixth Mass Extinction Already Arrived?" *Nature* 471 (7336): 51–57.

Beatty, John. 1995. "The Evolutionary Contingency Thesis." In *Concepts, Theories, and Rationality in the Biological Sciences*, edited by Gereon Wolters and James G. Lennox, 45–81. Pittsburgh: University of Pittsburgh Press.

Blank, G. D. 1988. "Metaphors in the Lexicon." *Metaphor and Symbolic Activity* 3: 21–36.

Blasko, D. G., and C. M. Connine. 1993. "Effects of Familiarity and Aptness on Metaphor Processing." *Journal of Experimental Psychology: Learning, Memory, and Cognition* 19 (2): 295–308.

Born, M., W. Heisenberg, and P. Jordan. 1926. "On quantum mechanics II." *Z. Phys* 35: 557–615.

Bowdle, Brian F., and Dedre Gentner. 2005. "The Career of Metaphor." *Psychological Review* 112 (1): 193.

Boyd, R. 1979. "Metaphor and Theory Change." In *Metaphor and Thought*, edited by A. Ortony, 356–408. Cambridge: Cambridge University Press.

Brandon, Robert N. 1990. *Adaptation and Environment*. Princeton, NJ: Princeton University Press.

Brown, Harvey R. 2005. *Physical Relativity: Space-Time Structure from a Dynamical Perspective*. New York: Oxford University Press.

Brush, S. G. 1992. "Alfven's Programme in Solar System Physics." *Plasma Science, IEEE Transactions on* 20 (6): 577–589.

Brush, Stephen G. 1996. "The Reception Mendeleev's Periodic Law in America and Britain." *Isis* 87 (4): 595–628.

Brush, Stephen G. 2009. *Choosing Selection: The Revival of Natural Selection in Anglo-American Evolutionary Biology, 1930–1970*. Philadelphia: American Philosophical Society.

Brush, Stephen G. 2015. *Making 20th Century Science: How Theories Became Knowledge*. Oxford: Oxford University Press.

Camp, Elisabeth. 2006. "Metaphor in the Mind: The Cognition of Metaphor 1." *Philosophy Compass* 1 (2): 154–170.

Camp, Elisabeth. 2009. "Two Varieties of Literary Imagination: Metaphor, Fiction, and Thought Experiments." *Midwest Studies In Philosophy* 33 (1): 107–130.

Camp, Elisabeth. 2019a. "Imaginative Frames for Scientific Inquiry: Metaphors, Telling Facts, and Just-So Stories." In *The Scientific Imagination*, edited by Arnon Levy and Peter Godfrey-Smith, 304–336. New York: Oxford University Press.

Camp, Elisabeth. 2019b. "Perspectives and Frames in Pursuit of Ultimate Understanding." In *Varieties of Understanding: New Perspectives from Philosophy, Psychology, and Theology*, edited by Stephen Grimm, 17–47. New York: Oxford University Press.

Cavalieri, Bonaventura. 1635. *Geometria indivisibilibus continuorum nova quadam ratione promota*. Bologna, Italy: Clemente Ferroni.

Chang, Hasok. 2022. *Realism for Realistic People: A New Pragmatist Philosophy of Science*. Cambridge: Cambridge University Press.

Cohen, I. Bernard. 1956. *Franklin and Newton; An Inquiry into Speculative Newtonian Experimental Science and Franklin's Work in Electricity as an Example Thereof*. Vol. 43. Memoirs of the American Philosophical Society. Philadelphia, PA: American Philosophical Society.

Cohen, I. Bernard. 1983. *The Newtonian Revolution*. New York: Cambridge University Press.

Collins, H. M. 1985. *Changing Order: Replication and Induction in Scientific Practice*. London: Sage.

Collins, Harry M. 2001. "Tacit Knowledge, Trust and the Q of Sapphire." *Social Studies of Science* 31 (1): 71–85.

Collins, John D. 2006. "Lotteries and the Close Shave Principle." In *Aspects of Knowing* dited by S. S. Hetherington, 83–96. Amsterdam: Elsevier.

Corsi, Pietro. 2005. "Before Darwin: Transformist Concepts in European Natural History." *Journal of the History of Biology* 38 (1): 67–83.

Curtius, Ernst Robert. 1953. *European Literature and the Latin Middle Ages*. Vol. 36. Bollingen series. New York: Pantheon Books.

Damerow, P., G. Freudenthal, P. McLaughlin, and J. Renn. 1991. *Exploring the Limits of Preclassical Mechanics: A Study of Conceptual Development in Early Modern Science: Free Fall and Compounded Motion in the Work of Descartes, Galileo and Beeckman*. New York: Springer.

Descartes, René. 1637. *The Geometry of Rene Descartes with a facsimile of the first edition*, translated by David E. Smith and Marcia L. Latham. New York: Dover Publications, Inc., 1954.

Drake, Stillman. 1978. *Galileo at Work: His Scientific Biography*. Chicago: University of Chicago Press.

Dummett, Michael. 1973. *Frege. Philosophy of Language*. London: Duckworth.

Einstein, Albert. 1920. *Relativity: The Special & the General Theory*. New York: Henry Holt and Company.

Elgin, Catherine Z. 2017. *True Enough*. Cambridge, MA: MIT Press.

Erwin, Douglas H., James W. Valentine, and J. John Sepkoski Jr. 1987. "A Comparative Study of Diversification Events: The Early Paleozoic Versus the Mesozoic." *Evolution* 41 (6): 1177–1186.

Fletcher, T. J. 1966. "Correspondence." *Mathematical Gazette* 50 (374): 398–401.

Frege, Gottlob. 1979. *Posthumous Writings*. Chicago: University of Chicago Press.

Frege, Gottlob. 1980. *The Foundations of Arithmetic: A Logico-Mathematical Enquiry into the Concept of Number*. 2nd rev. ed. Evanston, IL: Northwestern University Press.

Galison, Peter. 1987. *How Experiments End*. Chicago: University of Chicago Press.

Galison, Peter. 2016. "Practice All the Way Down." In *Kuhn's' Structure of Scientific Revolutions at Fifty*, edited by Robert Richards and Lorraine Daston, 42–70. Chicago: University of Chicago Press.

Gayon, Jean. 1998. *Darwinism's Struggle for Survival*. New York: Cambridge University Press.

Goodman, Nelson. 1954. *Fact, Fiction & Forecast*. London: University of London.

Goodman, Nelson. 1968. *Languages of Art; An Approach to a Theory of Symbols*. Indianapolis: Bobbs-Merrill.

Gould, Stephen Jay. 1966. "Allometry and Size in Ontogeny and Phylogeny." *Biological Reviews* 41 (4): 587–638.

Gould, Stephen Jay. 1970. "Dollo on Dollo's Law: Irreversibility and the Status of Evolutionary Laws." *Journal of the History of Biology* 3 (2): 189–212.

Gould, Stephen Jay. 1978. "Generality and Uniqueness in the History of Life: An Exploration with Random Models." *Bioscience* 28(4): 277–281.

Gould, Stephen Jay. 1980. "The Promise of Paleobiology as a Nomothetic, Evolutionary Discipline." *Paleobiology* 6 (1): 96–118.

Gould, Stephen Jay. 1984. "The Life and Work of TJM Schopf (1939–1984)." *Paleobiology* 10 (2): 280–285.

Gould, Stephen Jay. 1989. *Wonderful Life: The Burgess Shale and the Nature of History*. 1st ed. New York: W.W. Norton.

Gray, Jeremy. 2013. *Henri Poincaré: A Scientific Biography*. Princeton, NJ: Princeton University Press.

Greenberg, Daniel S. 2000. *Science, Money, and Politics*. Chicago: University of Chicago Press.

Grosholz, Emily. 2007. *Representation and Productive Ambiguity in Mathematics and the Sciences*. Oxford: Oxford University Press.

Hacking, I. 2012. "Introduction." In *The Structure of Scientific Revolutions*, edited by Thomas S. Kuhn, vii–xxxvii. Chicago: University of Chicago Press.

Hacking, Ian. 2016. "Paradigms." In *Kuhn's' Structure of Scientific Revolutions at Fifty*, edited by Robert Richards and Lorraine Daston, 96–114. Chicago: University of Chicago Press.

Hanson, Norwood Russell. 1958. *Patterns of Discovery; An Inquiry into the Conceptual Foundations of Science*. Cambridge: Cambridge University Press.

Haufe, Chris. 2015. "Gould's Laws." *Philosophy of Science* 82 (1): 1–20.

Haufe, Chris. 2022. *How Knowledge Grows*. Cambridge, MA: MIT Press.

Heideman, Michael T., Don H. Johnson, and C. Sidney Burrus. 1985. "Gauss and the History of the Fast Fourier Transform." *Archive for History of Exact Sciences* 34 (3): 265–277.

Heilbron, J. L. 1979. *Electricity in the 17th and 18th Centuries: A Study of Early Modern Physics*. Berkeley: University of California Press.

Hilbert, David. 1902. "Mathematical Problems." *Bulletin of the American Mathematical Society* 8 (10): 437–480.

Hobbes, Thomas. 1656. *Six Lessons to the Professors of Mathematiques, One of Geometry, the Other of Astronomy: In the Chaires Set Up by Sir Henry Savile in the University of Oxford*. London.

Hofmann, Josef Ehrenfried. 1939. "On the Discovery of the Logarithmic Series and Its Development in England up to Cotes." *National Mathematics Magazine* 14 (1): 37–45.

Holmes, Frederic Lawrence. 2004. *Investigative Pathways: Patterns and Stages in the Careers of Experimental Scientists*. New Haven, CT: Yale University Press.

Holton, Gerald. 1978. "Subelectrons, Presuppositions, and the Millikan-Ehrenhaft Dispute." *Historical Studies in the Physical Sciences* 9: 161–224.

Hubbell, Stephen P. 2005. "The Neutral Theory of Biodiversity and Biogeography and Stephen Jay Gould." *Paleobiology* 31 (2Suppl): 122–132.

Huss, John. 2009. "The Shape of Evolution: The MBL Model and Clade Shape." In *The Paleobiological Revolution: Essays on the Growth of Modern Paleontology*, edited by David Sepkoski and Michael Ruse, 327–345. Chicago: University Of Chicago Press.

Jablonski, David. 1986. "Background and Mass Extinctions: The Alternation of Macroevolutionary Regimes." *Science* 231 (4734): 129–133.

Jablonski, David. 2005. "Mass Extinctions and Macroevolution." *Paleobiology* 31 (2): 192–210.

Joy, Lynn S. 2006. "Scientific Explanation from Formal Causes to Law of Nature." In *The Cambridge History of Science*, edited by Katherine Park and Lorraine Daston, 70–105. New York: Cambridge University Press.

Kaiser, David. 2005. *Drawing Theories Apart: The Dispersion of Feynman Diagrams in Postwar Physics*. Chicago: University of Chicago Press.

Katz, Jerrold J. 1975. "Logic and Language: An Examination of Recent Criticism of Intensionalism." In *Language, Mind, and Knowledge*, edited by Keith Gunderson, 36–130. Minneapolis: University of Minnesota Press.

Kitcher, Philip. 1985. "Darwin's Achievement." In *Reason and Rationality in Natural Science*, edited by N. Rescher, 127–189. Lanham, MD: University Press of America.

Kitcher, Philip. 1989. "Explanatory Unification and the Causal Structure of the World." In *Scientific Explanation*, edited by Philip Kitcher and Wesley Salmon, 410–505. Minneapolis: University of Minnesota Press.

Kitcher, Philip. 1993. *The Advancement of Science.* New York: Oxford University Press.

Kitcher, Philip. 2001. *Science, Truth, and Democracy.* Oxford Studies in Philosophy of Science. Oxford: Oxford University Press.

Kline, Morris. 1972. *Mathematical Thought from Ancient to Modern Times.* New York: Oxford University Press.

Kohn, D. 1980. "Theories to Work by: Rejected Theories, Reproduction, and Darwin's Path to Natural Selection." *Studies in History of Biology* 4: 67–70.

Kipke, Saul. 1980. *Naming and Necessity.* Cambridge, MA: Harvard University Press.

Kuhn, Thomas S. 1977. "Objectivity, Value Judgement, and Theory Choice." In *The Essential Tension*, edited by T. S. Kuhn, 320–339. Chicago: University of Chicago Press.

Kuhn, Thomas S. 1993. "Afterwords." In *World Changes: Thomas Kuhn and the Nature of Science*, edited by Paul Horwich, 311–341. Cambridge, MA: MIT Press.

Kuhn, Thomas S. 1961. "The Function of Measurement in Modern Physical Science." *Isis* 52 (2): 161–193. http://www.jstor.org/stable/228678.

Kuhn, Thomas S. 1962/2012. *The Structure of Scientific Revolutions.* Chicago: University of Chicago Press.

Kuhn, Thomas S. 1970. *The Structure of Scientific Revolutions.* 2nd ed. Chicago: University of Chicago Press.

Kuhn, Thomas S. 1974. "Second Thoughts on Paradigms." In *The Structure of Scientific Theories*, edited by Frederick Suppe, 459–482. Urbana: University of Illinois Press.

Kuhn, Thomas S. 1979. "Metaphor in Science." In *Metaphor and Thought*, edited by Andrew Ortony, 533–542. Cambridge: Cambridge University Press. Reprinted in T.S. Kuhn 2000, 196–207.

Kuhn, Thomas S. 2000. *The Road Since Structure: Philosophical Essays, 1970-1993, with an Autobiographical Interview.* Chicago: University of Chicago Press.

Lange, Marc. 2011. "Why Do Forces Add Vectorially? A Forgotten Controversy in the Foundations of Classical Mechanics." *American Journal of Physics* 79 (4): 380–388.

Lange, Marc. 2014. "Aspects of Mathematical Explanation: Symmetry, Unity, and Salience." *Philosophical Review* 123 (4): 485–531.

Laudan, Larry. 1977. *Progress and Its Problems*. Berkeley: University of California Press.

Laudan, R. 1993. "Histories of the Sciences and Their Uses: A Review to 1913." *History of Science* 31(91): 1–34.

Laudan, Rachel, and Larry Laudan. 1989. "Dominance and the Disunity of Method: Solving the Problems of Innovation and Consensus." *Philosophy of Science* 56 (2): 221–237.

Lustig, A. J. 2008. "Darwin's Difficulties." In *The Cambridge Companion to the 'Origin of Species'*, edited by Michael Ruse and Robert J. Richards, 109–128. Cambridge: Cambridge University Press.

MacArthur, Robert H., and Edward O. Wilson. 1967. *The Theory of Island Biogeography*. Vol. 1. Monographs in Population Biology. Princeton, NJ: Princeton University Press.

Machamer, Peter. 1998. "Galileo's Machines, His Mathematics, and His Experiments." *The Cambridge Companion to Galileo*: 53–79.

Mahoney, Michael S. 1973. *The Mathematical Career of Pierre de Fermat (1601–1665)*. Princeton, NJ: Princeton University Press.

Mahoney, Michael Sean. 1998. "The Mathematical Realm of Nature." In *Cambridge History of Seventeenth-Century Philosophy*, edited by Daniel Garber and Michael Ayers, 702–755. New York: Cambridge University Press.

Maxwell, James Clerk, and William Davidson Niven. 1890. *The Scientific Papers of James Clerk Maxwell*. 2 vols. Cambridge: Cambridge University Press.

McAllister, J. W. 1999. *Beauty and Revolution in Science*. Ithaca: Cornell University Press.

McShea, D. W., and Robert N. Brandon. 2010. *Biology's First Law*. Chicago: University of Chicago Press.

Mercator, Nicholas. 1668. *Logarithmotechnica*. London: William Godbid and Moses Pitt.

Miller, A. 1987. "Symmetry and Imagery in the Physics of Bohr, Einstein, and Heisenberg." In 'Symmetries in Physics (1600–1980)', *Proceedings of the 1st International Meeting of the History of Scientific Ideas*, edited by Doncel, M. A. et al., 299–325. Bellaterra: Barcelona.

Moran, Richard. 1989. "Seeing and Believing: Metaphor, Image, and Force." *Critical Inquiry* 16 (1): 87–112.

Muntersbjorn, Madeline M. 2003. "Representational Innovation and Mathematical Ontology." *Synthese* 134 (1/2): 159–180.

Newton, Isaac, and D. T. Whiteside. 1967. *The Mathematical Papers of Isaac Newton*. Vol. 1. Cambridge: Cambridge University Press.

Newton, Isaac, and D. T. Whiteside. 1968. *The Mathematical Papers of Isaac Newton*. Vol. 2. Cambridge: Cambridge University Press.

Swerdlow, Noel M., and Neugebauer, Otto. 1984. *Mathematical Astronomy in Copernicus's* De Revolutionibus. Germany: Springer.

Nickles, Thomas. 2013. "Some Puzzles about Kuhn's Exemplars." In *Kuhn's The Structure of Scientific Revolutions Revisited*, edited by Vasso Kindi and Theodore Arabatzis, 112–133. New York: Taylor & Francis.

Ospovat, Dov. 1981. *The Development of Darwin's Theory*. Cambridge: Cambridge University Press.

Poincaré, Henri. 1910. "The Future of Mathematics." *The Monist* 20 (1): 76–92.

Poincaré, Henri. 1914. *Science and Method*. London: T. Nelson and sons.

Poincaré, Henri. 1946. *The Foundations of Science: Science and Hypothesis, the Value of Science, Science and Method*. Lancaster, PA: The Science Press.

Polanyi, Michael. 1958. *Personal Knowledge; Towards a Post-critical Philosophy*. Chicago: University of Chicago Press.

Priest, G. 1976. Gruesome Simplicity. *Philosophy of Science* 43 (3): 432–437.

Putnam, Hilary. 1962. "It Ain't Necessarily So." *The Journal of Philosophy* 59 (22): 658–671.

Quine, W. V. 1951. "Two Dogmas of Empiricism." *The Philosophical Review* 60 (1): 20–43.

Raup, David M. 1991. *Extinction: Bad Genes or Bad Luck?* New York: W.W. Norton.

Raup, David M., and Stephen Jay Gould. 1974. "Stochastic Simulation and Evolution of Morphology—Towards a Nomothetic Paleontology." *Systematic Zoology* 23 (3): 305–322.

Rehbock, Philip F. 1983. *The Philosophical Naturalists: Themes in Early Nineteenth-Century British Biology*. Madison: University of Wisconsin Press.

Reichenbach, Hans. 1938. *Experience and Prediction; An Analysis of the Foundations and the Structure of Knowledge*. Chicago: University of Chicago Press.

Rheinberger, Hans-Jörg. 1997. *Toward a History of Epistemic Things: Synthesizing Proteins in the Test Tube*. Writing Science. Stanford, CA: Stanford University Press.

Rocke, Alan J. 2010. *Image and Reality: Kekulé, Kopp, and the Scientific Imagination*. Chicago: The University of Chicago Press.

Roller, Duane, and Duane H. D. Roller. 1950. "The Development of the Concept of Electric Charge." In *Harvard Case Studies in Experimental Science*, edited by James Conant, 1–93. Cambridge, MA: Harvard University Press.

Rothman, Aviva. 2017. *The Pursuit of Harmony: Kepler on Cosmos, Confession, and Community*. Chicago: The University of Chicago Press.

Rudwick, M. J. S. 1972. *The Meaning of Fossils: Episodes in the History of Palaeontology*. New York: Macdonald and Co.

Ruse, Michael. 1979. *The Darwinian Revolution: Science Red in Tooth and Claw*. Chicago: University of Chicago Press.

Sepkoski, David. 2012. *Rereading the Fossil Record*. Chicago: University of Chicago Press.

Sepkoski, David, and Michael Ruse. 2009. *The Paleobiological Revolution: Essays on the Growth of Modern Paleontology*. Chicago: University of Chicago Press.

Sepkoski, Jr, J. John. 1978. "A Kinetic Model of Phanerozoic Taxonomic Diversity I. Analysis of Marine Orders." *Paleobiology* 4 (3): 223–251.

Sepkoski, Jr, J. John. 1979. "A Kinetic Model of Phanerozoic Taxonomic Diversity II. Early Phanerozoic Families and Multiple Equilibria." *Paleobiology* 5 (3): 222–251.

Sepkoski, Jr, J. John. 1984. "A Kinetic Model of Phanerozoic Taxonomic Diversity. III. Post-Paleozoic Families and Mass Extinctions." *Paleobiology* 10 (2): 246–267.

Shapin, Steven. 2010. *Never Pure*. Baltimore: The Johns Hopkins University Press.

Shapin, Steven and Simon Schaffer. 1985. *Leviathan and the Air-Pump: Hobbes, Boyle, and the Experimental Life: Including a Translation of Thomas Hobbes, Dialogus physicus de natura aeris by Simon Schaffer*. Princeton, NJ: Princeton University Press.

Shimony, A. 1989. "The Non-existence of a Principle of Natural Selection." *Biology and Philosophy* 4 (3): 255–273.

Sholl, D. A. 1954. "Regularities in Growth Curves, Including Rhythms and Allometry." In *Dynamics of Growth Processes*, edited by E. J. Boell, 224–241. Princeton: Princeton University Press.

Simon, Herbert A. 1973. "The Structure of Ill Structured Problems." *Artificial Intelligence* 4 (3): 181–201.

Slater, M. H. 2015. "Natural Kindness." *The British Journal for the Philosophy of Science* 66 (2): 375–411.

Slotten, R. A. 2004. *The Heretic in Darwin's Court: The Life of Alfred Russel Wallace*. New York: Columbia University Press.

Stedall, Jacqueline. 2005. "John Wallis, Arithmetica infinitorum (1965)." In *Landmark Writings in Western Mathematics 1640–1940*, edited by I. Grattan-Guinness, Roger Cooke, Leo Corry, Pierre Crépel, and Niccolo Guicciardini, 23–32. Amsterdam: Elsevier Science.

Swerdlow, Noel Mark, and Otto Neugebauer. 1984. *Mathematical Astronomy in Copernicus's De revolutionibus*. 2 vols. Vol. 10. Studies in the History of Mathematics and Physical Sciences. New York: Springer-Verlag.

Swerdlow, Noel Mark. 2012. "Copernicus and Astrology, with an Appendix of Translations of Primary Sources." *Perspectives on Science* 20 (3): 353–378.

't Hooft, Gerardus. 1971. "Renormalization of Massless Yang-Mills Fields." *Nuclear Physics: B* 33 (1): 173–199.

't Hooft, Gerard. 2016. "Reflections on the Renormalization Procedure for Gauge Theories." *Nuclear Physics B* 912: 4–14.

Tappenden, Jamie. 1995. "Extending Knowledge and Fruitful Concepts: Fregean Themes in the Foundations of Mathematics." *Noûs* 29 (4): 427–467.

Tappenden, Jamie. 2008. "Mathematical Concepts: Fruitfulness and Naturalness." In *The Philosophy of Mathematical Practice*, edited by Paolo Mancosu, 276–301. New York: Oxford University Press.

Tappenden, Jamie. 2012. "Fruitfulness as a Theme in the Philosophy of Mathematics." *The Journal of Philosophy* 109 (1/2): 204–219.

Tappenden, Jamie. 2018. "Frege, Karl Snell and Romanticism; Fruitful Concepts and the 'Organic/Mechanical' Distinction." Unpublished.

Thomson, J. J. 1907. *The Corpuscular Theory of Matter*. 2nd ed. London: A. Constable & Co.

Todes, Daniel. 1989. *Darwin Without Malthus*. New York: Oxford University Press.

Todes, Daniel. 2009. "Global Darwin: Contempt for Competition." *Nature* 462 (7269): 36–37.

Torricelli, Evangelista. 1644. *Opera geometrica*. Florence: Landi e Massa.

Truesdell, Clifford. 1968. *Essays in the History of Mechanics*. Berlin: Springer-Verlag.

Truesdell, Clifford. 1967. "Reactions of Late Baroque Mechanics to Success, Conjecture, Error, and Failure in Newton's Principia." *The Texas Quarterly* 10 (3): 238–258.

Wallis, John. 1656. *Arithmetica Infinitorum*. Oxford.

Wallis, John. 1685. *A Treatise of Algebra Both Historical and Practical*. London: R. Davis.

Walter, Scott. 2009. "Hypothesis and Convention in Poincarés' Defense of Galilei Spacetime." In *The Significance of the Hypothetical in the Natural Sciences*, edited by Michael Heidelberger and Gregor Schiemann, 193–219. Walter de Gruyter.

Weinberg, Steven. 1967. "A Model of Leptons." *Physical Review Letters* 19 (21): 1264–1266.

Weinberg, Steven. 1980. "Conceptual Foundations of the Unified Theory of Weak and Electromagnetic Interactions." *Science* 210 (4475): 1212–1218.

Weisberg, Michael, and Ryan Muldoon. 2009. "Epistemic Landscapes and the Division of Cognitive Labor." *Philosophy of Science* 76 (2): 225–252.

Westfall, Richard S. 1971. *Force in Newton's Physics: The Science of Dynamics in the Seventeenth Century*. London: Macdonald and Co.

Westfall, Richard S. 1977. *The Construction of Modern Science: Mechanisms and Mechanics*. New York: Cambridge University Press.

Westfall, Richard S. 1980. *Never at Rest: A Biography of Isaac Newton*. Cambridge: Cambridge University Press.

Whiteside, Derek Thomas. 1961. "Patterns of Mathematical Thought in the Later Seventeenth Century." *Archive for History of Exact Sciences* 1: 179–388.

Wittgenstein, Ludwig. 1953. *Philosophical Investigations*. Oxford: B. Blackwell.

Wray, K. Brad. 2011. *Kuhn's Evolutionary Social Epistemology*. New York: Cambridge University Press.

Wray, K. Brad. 2021. *Kuhn's Intellectual Path: Charting the Structure of Scientific Revolutions*. New York: Cambridge University Press.

Zollman, Kevin James Spears. 2010. "Social Structure and the Effects of Conformity." *Synthese* 172 (3): 317–340.

Index

For the benefit of digital users, indexed terms that span two pages (e.g., 52–53) may, on occasion, appear on only one of those pages.

2 04